Evgeniya Ballmann

Physics-Based Probabilistic Motion Compensation of Elastically Deformable Objects

Karlsruhe Series on Intelligent Sensor-Actuator-Systems

Volume 10

ISAS | Karlsruhe Institute of Technology
Intelligent Sensor-Actuator-Systems Laboratory

Edited by Prof. Dr.-Ing. Uwe D. Hanebeck

Physics-Based Probabilistic Motion Compensation of Elastically Deformable Objects

by

Evgeniya Ballmann

Dissertation, Karlsruher Institut für Technologie (KIT)
Fakultät für Informatik
Tag der mündlichen Prüfung: 10. Februar 2012

Impressum

Karlsruher Institut für Technologie (KIT)
KIT Scientific Publishing
Straße am Forum 2
D-76131 Karlsruhe
www.ksp.kit.edu

KIT – Universität des Landes Baden-Württemberg und
nationales Forschungszentrum in der Helmholtz-Gemeinschaft

KIT Scientific Publishing 2012
Print on Demand

ISSN 1867-3813
ISBN 978-3-86644-862-9

Physics-Based Probabilistic Motion Compensation of Elastically Deformable Objects

zur Erlangung des akademischen Grades eines

Doktors der Ingenieurwissenschaften

von der Fakultät für Informatik
des Karlsruher Instituts für Technologie (KIT)

genehmigte

Dissertation

von

Evgeniya Ballmann

Tag der mündlichen Prüfung: 10.02.2012

Erster Gutachter: Prof. Dr.-Ing. Uwe D. Hanebeck

Zweiter Gutachter: Prof. Dr.-Ing. Tobias Ortmaier

Acknowledgements

This thesis was written at the Intelligent Sensor-Actuator-Systems Laboratory (ISAS) of Karlsruhe Institute of Technology (KIT). I have collaborated with many people from different disciplines within the Research Training Group 1126 "Intelligent Surgery" of the German Research Foundation, who I am grateful for encouraging me in various ways.

In particular, I would like to thank Prof. Dr.-Ing. Uwe D. Hanebeck for advising me, fruitful discussions, and sharing his fascination for probabilistic theory. My sincere gratitude goes to Prof. Dr.-Ing. Tobias Ortmaier for reviewing this thesis. I am also grateful to Prof. Dr. med. Raffaele de Simone, Prof. Dr. med. Gábor B. Szabó, and Dr. med. Tobias Gehrig from the Heidelberg University Hospital for the extensive support in medical issues.

Many sincere thanks are extended to my colleagues at ISAS. Without them, the three and a half years of my research work would not have been so enjoyable. I would especially like to express my great appreciation to Vesa Klumpp and Frederik Beutler for many thought-provoking conversations. To Benjamin Noack, Florian Faion, and Patrick Ruoff, thanks for your friendship and support. Our dinner discussions about political and social issues were an unforgettable respite from the research.

My special thanks go to my colleagues at the medical group of the Humanoids and Intelligence Systems Lab of the KIT. Most of all, I owe a debt of gratitude to Stefanie Speidel for many stimulating talks, as well as to Sebastian Röhl and Stefan Suwelack, with whom I puzzled over many of the same problems in physical modeling and image processing.

Furthermore, I would like to offer my sincere gratitude to the students, who worked with me at various aspects of this thesis, especially to Andreas Hofmann, Pascal Pompey, and Arne Schumann for their great enthusiasm and extraordinary commitment.

Moreover, my very great appreciation goes to my friends and my family for their encouragements and an enormous support.

Karlsruhe, August 2012
Evgeniya Ballmann

Contents

Notations

General Conventions

$\mathbb{Z}, \mathbb{Z}_+$	Set of integer/non-negative integer numbers
$\mathbb{R}, \mathbb{R}_+$	Set of real/non-negative real numbers
$x, \underline{x}$	Scalar/vector
$x_k, \underline{x}_k$	Scalar/vector at time step t_k
$\underline{x}_{1:k}$	Time sequence of vectors $\{\underline{x}_1, \underline{x}_2 \ldots, \underline{x}_k\}$
$\boldsymbol{x}, \underline{\boldsymbol{x}}$	Random variable/vector
$\hat{\underline{x}}$	Known value of random vector
x^r, y^r, z^r	Components of the vector $\underline{r} = \left[x^r, y^r, z^r\right]^{\mathrm{T}}$
n_x	Dimension of the vector $\underline{x}$
$\mathbf{A}$	Matrices are denoted by bold upper case letters
$\mathbf{I}$	Identity matrix
$\mathbf{0}$	Zero matrix
$\mathcal{P}$	Sets are denoted by calligraphic upper case letters
$\boldsymbol{\mathcal{P}}$	Random sets are denoted by bold calligraphic upper case letters
$\mathcal{P}_k$	Set at time step t_k
$\mathcal{P}_{1:k}$	Time sequence of sets $\{\mathcal{P}_1, \mathcal{P}_2, \ldots, \mathcal{P}_k\}$
N_P	Number of elements in the set $\mathcal{P}$
$\mathbb{W}$	Functionals are denoted by mathematical double-struck upper case letters
$\|\cdot\|$	Norm in a vector space
$\sim$	Distribution operator, e.g., $\underline{\boldsymbol{x}} \sim f(\underline{x})$ denotes that a random vector $\underline{\boldsymbol{x}}$ is distributed according to the probability density function f
$\approx$	Approximation operator, e.g., $x \approx y$ denotes that x is approximated by y
$\mapsto$	Functional mapping of vectors, e.g., $\underline{x} \mapsto \underline{y}$ means that the vector $\underline{x}$ maps to the vector $\underline{y}$
$\rightarrow$	Mapping of sets, e.g., $\mathcal{P}_1 \rightarrow \mathcal{P}_2$ means that $\mathcal{P}_1$ maps to $\mathcal{P}_2$

Function Symbols

$\mathbf{A}^{+}$	Pseudo-inverse of the matrix $\mathbf{A}$
$(\mathbf{A})^{-1}$	Inverse of the matrix $\mathbf{A}$
$\underline{x}^{\mathrm{T}}, \mathbf{A}^{\mathrm{T}}$	Vector $\underline{x}$/matrix $\mathbf{A}$ transpose
$\underline{\breve{x}}, \breve{\mathbf{A}}, \breve{\mathcal{P}}$	Vector $\underline{x}$, matrix $\mathbf{A}$, and set $\mathcal{P}$ enlarged by additional elements
$\operatorname{trace}(\mathbf{A})$	Trace of the matrix $\mathbf{A}$
$\operatorname{div}\mathbf{A}$	Divergence of the matrix $\mathbf{A}$
$\operatorname{diag}\{\underline{x}\}$	Diagonal matrix with elements of the vector $\underline{x}$ on main diagonal
$\mathbf{A} : \mathbf{B}$	Frobenius inner product of matrices $\mathbf{A}$ and $\mathbf{B}$
$\underline{\dot{x}}, \underline{\ddot{x}}$	First/second time derivative of the vector function $\underline{x}$
$\frac{\partial y}{\partial x}$	Partial derivative of function y with respect to the variable x
$\min f(x), \max f(x)$	Minimum/maximum of the function f
$\underset{x}{\arg\min}\, f(x)$	Argument x that minimizes the function f
O	Big-O in Landau notation, e.g., $f(x) \in O(g(x))$ means that the function f has the order of $g(x)$

Symbols for Probability Densities

$f(\underline{x})$	General probability density of $\underline{x}$
$f(\underline{x}, \underline{y})$	Joint probability density of $\underline{x}$ and $\underline{y}$
$f(\underline{x} \mid \underline{y})$	Conditional probability density, e.g., conditioning the random vector $\underline{x}$ on the vector $\underline{y}$
$f_k^{\mathrm{p}}(\underline{x}_k)$	A priori density at time step t_k
$f_k^{\mathrm{e}}(\underline{x}_k)$	A posteriori density after measurement update at time step t_k
$f_k^{\mathrm{T}}(\underline{x}_k \mid \underline{x}_{k-1})$	Transition density at time step t_k
$f_k^{\mathrm{L}}(\hat{\underline{y}}_k \mid \underline{x}_k)$	Likelihood at time step t_k
$\mathcal{N}\left(\underline{x}_k, \underline{\mu}_k^x, \mathbf{\Sigma}_k^x\right)$	Gaussian density of vector $\underline{x}_k$ with mean vector $\underline{\mu}_k^x$ and covariance matrix $\mathbf{\Sigma}_k^x$
$\delta(\underline{x})$	Dirac delta distribution

Symbols for Continuous and Semi-Discrete Models

$\underline{d}$	Displacement field in vector form
$\underline{c}, \underline{m}, \underline{r}$	Vector of nodal values/model nodes/collocation points
$\boldsymbol{\mathcal{E}}, \underline{\epsilon}$	Cauchy's strain tensor in matrix/vector form
$\boldsymbol{\Sigma}, \underline{\sigma}$	Stress tensor in matrix/vector form
$\underline{f}_N^u, \underline{f}_N^n$	Vectors of uniform and non-uniform pressure forces
$\underline{d}^h, \underline{\epsilon}^h, \underline{\sigma}^h$	Approximated displacement, strain, and stress fields
$\psi, \underline{\psi}$	Shape function and vector of shape functions
Ω	Bounded domain of model in mechanical equilibrium
Ω^ψ	Bounded domain of model out of mechanical equilibrium
$\overline{\Omega}, \overline{\Omega}^\psi$	Closures of the domain Ω/Ω^ψ
$\partial\Omega, \partial\Omega^\psi$	Boundaries of the domain Ω/Ω^ψ
$\Gamma_\mathcal{N}, \Gamma_\mathcal{R}$	Neumann/Robin boundaries
$\mathcal{M}, \check{\mathcal{S}}$	Set of model nodes $\underline{m}^j$/additional model nodes $\underline{\check{s}}^j$
$\mathbf{C}$	Material matrix
$\mathbf{M}, \mathbf{M}^{ij}$	Mass matrix and its individual entries
$\mathbf{V}, \mathbf{V}^{ij}$	Damping matrix and its individual entries
$\mathbf{K}, \mathbf{K}^{ij}$	Stiffness matrix and its individual entries
$\boldsymbol{\Phi}, \boldsymbol{\Phi}^{ij}$	Shape matrix and its individual entries
$\mathbf{A}, \mathbf{B}$	System and input matrices of a time-continuous system

Symbols for Discrete Models

$\underline{a}_k, \underline{h}_k$	System/measurement function of a discrete time system
$\underline{z}_k$	System state at time step t_k
$\underline{\xi}_k$	Augmented state vector at time step t_k
$\underline{\hat{u}}_k, \underline{u}_k$	Known and unknown input vector
$\underline{s}_k, \underline{e}_k$	Systematic errors of the system and measurement model
$\underline{w}_k^z$	Noise of the random vector $\underline{z}_k$ at time step t_k
$\underline{v}_k$	Measurement noise vector
$\underline{\theta}_k$	Parameter vector
$\mathbf{A}_k, \mathbf{B}_k$	System and input matrices of a discrete time system
$\mathbf{H}_k$	Measurement matrix of a discrete time system
$\underline{x}_k^l, \underline{x}_k^n$	Linear/nonlinear sub-vector of the augmented state vector $\underline{\xi}_k$
$\underline{s}_k^p, \underline{s}_k^e$	A priori/a posteriori estimate of the vector $\underline{s}_k$ at time step t_k

Symbols for Image Processing

$\underline{l}_k^i,$	Vector denoting position of ith landmark at time step t_k
$\underline{s}_k^i$	Positions of ith surface point at time step t_k
$\underline{f}_k^{j,i}$	Positions of ith image feature on the image of jth camera
$\underline{p}_k^{j,i}$	Position of ith pixel on the image of jth camera
$\underline{\hat{y}}_k^{j,i}$	Measurement of ith point provided by jth camera
$\underline{l}_k^{*i}$	Triangulated position of the landmark $\underline{l}^i$
$\underline{f}_k^{*j,i}$	Approximated position of the image feature $\underline{f}_k^{j,i}$ defined by an intersection of epipolar lines
$\underline{\varphi}^l, \underline{\varphi}^f$	Function that maps a vector in $\mathbb{R}^3/\mathbb{R}^2$ from Cartesian into homogeneous coordinate system
$\underline{\phi}^l, \underline{\phi}^f$	Functions that map a vector in $\mathbb{R}^3/\mathbb{R}^2$ from homogeneous into Cartesian coordinate system
I_k	Color function
T_k^j	Transformation function of jth camera
I_E, I_T, I_C	Indicator functions of epipolar, triangulation, and consistency criteria
C_P	Function defining the physics-based criterion
$\mathbf{D}_k$	Assignment matrix at time step t_k
$\mathbf{P}^j$	Projection matrix of jth camera
$\mathbf{R}^j$	Matrix of extrinsic parameters of jth camera
$\mathbf{K}^j$	Matrix of intrinsic parameters of jth camera
$\underline{t}^j$	Translation vector of jth camera
$\mathbf{F}_{ij}$	Fundamental matrix of transformation from jth image to ith image
$\mathcal{P}_k^j, \mathcal{R}_k^j$	Sequences of pixels in the original/stabilized image of jth camera
$\mathcal{F}_k^j, \mathcal{Y}_k^j$	Sequences of image features/measurements on the image of jth camera
$\mathcal{L}_k, \mathcal{S}_k$	Sequences of landmarks/surface points
$\mathcal{P}_{1:k}^j, \mathcal{R}_{1:k}^j$	Time sequences of sets denoting original/stabilized images of jth camera
$\mathcal{P}_{1:k}, \mathcal{R}_{1:k}$	Unions of the sequences $\mathcal{P}_{1:k}^j/\mathcal{R}_{1:k}^j$ of all cameras
$a_k^{t^j}, a_k^{s^{j,i}}$	Average transformation/stabilization errors
a_k^c	Average motion compensation error

Constants and Units

ρ	in kg/m^3	Material density
E	in N/m^2	Modulus of elasticity
ν	-	Poisson ratio
λ	-	First Lamé constant
μ	in N/m^2	Second Lamé constant
β	in N/m	Stiffness of material on the Robin boundary
γ	in s	Rayleigh coefficient of damping proportional to the material stiffness
κ	in s^{-1}	Rayleigh coefficient of damping proportional to the mass
γ_E	in pixel	Gating parameter of epipolar criterion
γ_P	in units of standard deviation	Gating parameter of physics-based criterion
k	in pixel	Gating parameter for system refinement

Glossary

SPDE	Stochastic partial differential equations
SODE	Stochastic ordinary differential equations

Zusammenfassung

Die Bewegungskompensation elastisch deformierbarer Objekte ist in vielen industriellen und medizinischen Anwendungen von hoher Relevanz. Das betrifft beispielsweise die robotergestützten chirurgischen Eingriffe bei einer Operation am schlagenden Herz. Hier wird die Herzoberfläche mit einem Multikamerasystem verfolgt, um den Roboter mit der Herzbewegung zu synchronisieren. Gleichzeitig wird eine visuelle Bewegungskompensation für den Chirurgen durchgeführt, bei der die zeitlich und räumlich veränderliche Herzoberfläche für ihn als stillstehend dargestellt wird.

Im Gegensatz zu den gängigen Methoden, welche die Bewegung elastischer Objekte nur an ausgewählten Punkten kompensieren, liegt der Fokus dieser Arbeit in der Betrachtung der Deformation des gesamten Objekts. Aufgrund verrauschter Messdaten und ständiger Änderung des Bildinhalts stellt allein schon die Bewegungskompensation an den Messpunkten eine Herausforderung dar. Hinzu kommt, dass auch die Bewegung zwischen den Messpunkten kompensiert werden muss. Die Problematik verschärft sich, wenn das Objekt für einen langen Zeitraum verdeckt ist, so dass keine Messinformation über seine Verformung verfügbar ist.

Die zwei zentralen Beiträge dieser Arbeit sind das prädiktive Trackingverfahren und eine neue Methode für die visuelle Bewegungskompensation. Im Kontrast zu nicht modellbasierten Verfahren, die beispielsweise auf der reinen Interpolation zwischen Messpunkten beruhen, wird hier ein physikalisches Modell der Herzwand eingeführt, das die physikalischen Eigenschaften des realen Objekts in das Tracking und die Bewegungskompensation einbezieht. Dieses Modell dient als Grundlage für die mathematischen Modelle beider Verfahren. Ein wichtiger Aspekt der Modellierung ist die Balance zwischen Einfachheit und Genauigkeit. Zu diesem Zweck werden die vereinfachten Modelle mit einer detaillierten Beschreibung ihrer stochastischen und systematischen Fehler kombiniert. Ihre hohe Genauigkeit wird vor allem durch eine adaptive Anpassung bestimmt. Dies ermöglicht, die Raumdiskretisierung und die physikalischen Eigenschaften der Modelle nur da zu detaillieren, wo es notwendig ist.

Um eine hinreichend gute Näherung der Herzbewegung zu erreichen, wird die Herzwand mit einem linear viskoelastisch deformierbaren Volumenkörper modelliert, der die Bewegung auch im Inneren der Herzwand abbildet. Mathematisch wird das Modell durch ein System stochastischer partieller Differentialgleichungen mit unsicheren Eingangsdaten, sowie unsicheren Anfangs- und Randbedingungen beschrieben. Zur Schätzung der Herzbewegung muss dieses System in ein entsprechendes Zustandsmodell konvertiert werden. Da es nicht analytisch lösbar ist, wird dazu die Linienmethode verwendet, die das System zuerst im Raum und anschließend in der Zeit diskretisiert. Für die Raumdiskretisierung wird die elementfreie Methode benutzt, die die Kollokationsmethode mit dem Petrov-Galerkin-Verfahren vereint. Diese Methode zeichnet sich durch eine punktbasierte Diskretisierung des Modells aus, wobei keine vorab definierte Verbindung zwischen den Diskretisierungspunkten notwendig ist. Im Vergleich zu klassischen gitterbasierten Methoden, wie beispielsweise die Methode der finiten Elemente, die das Modell mit Elementen diskretisiert, ist die elementfreie Methode effizienter und flexibler. Dies kommt besonders zum Tragen bei der Verfeinerung der Raumdiskretisierung, da zusätzliche Diskretisierungspunkte leicht eingefügt oder entfernt werden können.

Das prädiktive Trackingverfahren bestimmt die wahrscheinlichen Positionen der Messpunkte auf physikalisch korrekte Weise. Da die physikalischen Eigenschaften der Herzwand unbekannt sind, wird zu diesem Zweck eine simultane Zustands- und Parameterschätzung durchgeführt. Als Folge der Verfeinerung der Modelle verändern sich die Dimensionen des Zustands und des Zustandsmodells. Zur Extraktion der Messinformation werden die wahrscheinlichen Positionen der Messpunkte einbezogen. Dies erlaubt die Messausreißer herauszufiltern. Zudem kann die dreidimensionale Bewegung des elastischen Objekts im Gegensatz zu nicht modellbasierten Trackingverfahren über längere Zeit auch bei vollständigem Verlust der Messinformation verfolgt werden.

Die visuelle Bewegungskompensation erstellt für die Bildsequenzen des Kamerasystems dazugehörige bewegungskompensierte Bildsequenzen. Die Bildtransformation wird mittels einer physikalisch-basierten Transformationsfunktion durchgeführt. Hergeleitet durch die Projektion der geschätzten Position des elastischen Objekts in die Kamerabilder, stellt diese Funktion die Korrespondenzen zwischen den Pixeln der Kamerabilder über die Zeit auf. Infolgedessen lassen sich die Bildänderungen durch Farbtransfer zwischen den zugeordneten Pixeln ausgleichen. Darüber hinaus ermöglicht die Rückkopplung der Kompensationsfehler die Qualität aller Modelle und somit des Trackings und der Bewegungskompensation kontinuierlich zu verbessern.

In Bezug auf die Anwendung für robotergestützte chirurgische Operationen am schlagenden Herz werden die Verfahren durch zahlreiche Experimente an einem künstlichen schlagenden Herz umfassend validiert. Der Fehler der Bewegungskompensation reduziert sich um 57% im Vergleich zur deterministischen und rein geometrischen Bildtransformation. Die vorgestellten Methoden gehen jedoch über diese Anwendung hinaus und können auch in anderen Bereichen, beispielsweise bei der robotergestützten Montage elastischer Objekte oder in der Videoverarbeitung, eingesetzt werden.

Abstract

Motion compensation of elastically deformable objects is of high importance for many industrial and medical applications. For example, it concerns the computer-assisted surgery system for operations on a beating heart, during which it is observed by a multi-camera system in order to enable the synchronization of the surgical robot with the heart surface. The spatially and temporally varying beating heart is represented to the surgeon as being motionless by means of visual motion compensation.

In contrast to standard methods that compensate the motion of elastic objects only at selected points, the focus of this work is in considering the deformation of the entire object. Motion compensation on measurement points alone is already challenging because of the continuous motion of the heart and noisy measurements. Compensating the motion between the measurement points further complicates the problem. This issue is aggravated when the object is hidden by obstacles over a long time horizon, during which no measurement information about its motion is available.

The two main contributions of this work are a predictive tracking approach and a novel method for visual motion compensation. In contrast to non-model-based methods exploiting, for example, a pure interpolation between the measurement points, a novel physical model of the heart wall builds the core of the proposed approaches. This enables the incorporation of the physical properties of this object into the motion tracking and compensation system. All mathematical models used by this system are derived from the physical model. An important aspect of modeling is the balance between its complexity and its accuracy. For this purpose, simplified models are combined with the detailed description of their stochastic and systematic errors. An adaptive refinement of the spatial discretization and the physical properties allows to add details where necessary, thus efficiently yielding a high accuracy of the models.

The heart wall is modeled as a linear deformable object with viscoelastic properties, in order to achieve a good approximation of the heart behavior. Due to its volumetric nature, it is able to reproduce the motion also in the interior of the heart wall. Mathematically formulated by a system of stochastic partial differential equations with an uncertain input and uncertain initial and boundary conditions, this system should be converted into an appropriate state-space model for the task of estimating the heart motion. Not analytically solvable, this system is discretized according to method of lines, first in space and then in time. For spatial discretization, the element-free method, specially the meshless local Petrov-Galerkin mixed collocation method, is applied. This method is characterized by a point-based discretization of the spatial domain, where no predefined connection between the discretization points is needed. It is more efficient and flexible than classical grid-based methods, such as finite element method that discretizes the spatial domain by elements. In addition, the refinement of the spatial discretization is particularly advantageous thanks to ease in adding or removing of discretization points.

The predictive tracking approach estimates the most probable positions of measurement points. As the physical properties of the heart are unknown, the simultaneous state and parameter estimation is employed for this task. As a result of the models refinement, the dimensions of the state and of the state-space model are continuously changing. The predicted positions of the measurement points determined in a physically correct way are involved for extracting measurement information. This enables filtering out measurement outliers. Furthermore, the three-dimensional motion of the object under observation can be tracked over a long time horizon even for total loss of measurement information, which is in contrast to non-model-based tracking methods.

The visual motion compensation transforms the image sequences provided by a camera system into stabilized image sequences. This transformation is based upon a physics-based transformation function. Derived by projecting the estimated position of the object onto the camera images, this function establishes the correspondences between the pixels of the camera images over time. Then, the changes in the images are compensated by color transfer between the corresponding pixels. Ultimately, the error feedback enables the continuous improvement of the quality of all models and therefore, of the tracking and visual motion compensation.

With respect to application in a computer-assisted surgery system, the approaches are evaluated by various experiments on the artificial beating heart. As a result, the error of the motion compensation is decreased by 57% compared to the deterministic and pure geometric image transformation. The methods developed here are not limited to the considered application and can also be employed in other areas, such as, for example, the robot-based assembly of elastic objects or video processing.

1 Introduction

Motion compensation of elastically deformable objects is highly relevant
to a wide range of applications. For instance, it is essential in industrial
tasks associated with control of flexible manipulators [208, 211, 233], automated handling or assembly of elastic objects, such as an insertion of
a flexible wire into a hole [149] and utilization in presence of non-rigid
motion [111, 228]. With respect to medical applications, e.g., in radiotherapy [172] and beating heart surgery [150], the motion compensation is of
particular importance for computer-assisted systems that encompass tasks
of synchronization of surgical instruments with the moving tissues [229],
virtual representation of the continuously deforming organs as being motionless, and image-guided navigation [97, 213]. When applied to video
processing, the motion compensation handles image frames for video com-
pression and stabilization [112], registration of dynamic textures [85, 227],
facial animation [22], and tracking of the objects [129].

Within these applications, the objective of the motion compensation is
twofold. On one side, it predicts how the moving object, such as a vibrating
flexible manipulator, wire, or beating heart, should be deformed to remain
motionless, and with respect to video processing, how the images of the
moving scene should be modified to stay stationary. On the other side, it
uses the predicted deformation of the moving object for synchronization.
For example, this is the case in beating heart surgery, where the surgical
instruments move synchronously with the beating heart so that the relative
motion between them and the heart is compensated.

Since the motion compensation is only as accurate as the motion estimates,
accurate methods for tracking and reconstruction of elastically deformable
objects are necessary. Besides solving the problem of extracting the desired information from unreliable and noise-contaminated measurements,
these methods should estimate the motion of the entire object, even between sparse spatially distributed measurement points. In this context,
with an increasing number of measurement points the accuracy of the estimation can be significantly improved. Unfortunately, this is accompanied
by a rising computational complexity and is limited by the performance
of the measurement systems. Furthermore, the problem is exacerbated

when measurements fail, e.g., as a result of occlusion of the elastic object observed by the camera.

Ultimately, if the measurement uncertainties are high and a complete loss of measurement information may occur over a long-time horizon, the underlying physical information should be incorporated in the reconstruction of the object's motion. In this case, the methods – whether without explicit models or involving geometric or statistical models – bear inherent drawbacks such as sensitivity to measurement outliers or physically incorrect motion reconstruction resulting from not taking this information into consideration.

To meet these challenges within the scope of this thesis, the central focus is concentrated upon the physics-based motion compensation, which is based on the physical model operating directly in the physical space.

1.1 Application

The methods for motion compensation proposed in this thesis are mainly intended to be applied in a computer-assisted beating heart surgery system first proposed in [150]. Nevertheless, it is worth mentioning that they are not limited to this system and are expected to apply also to the other afore-mentioned industrial and medical applications as well as to video processing.

Beating Heart Surgery In comparison to traditional surgery, using a heart-lung machine for temporal stopping of the heart, the beating heart surgery offers many advantages for the patients, including lower mortality rate in hospitalization, significantly less pronounced inflammatory reactions and shorter postoperative treatment [52]. Furthermore, the patient clearly loses less blood during the operation and the need for blood transfusion and anti-fibrillation is extremely reduced [48]. Moreover, the side effects or post-operative complications, in particular, atrial fibrillation, kidney failure, severe gastrointestinal complications, or neurocognitive dysfunction, occur rarely. As a result, the hospital benefits from significantly lower costs on surgical equipment and post-operative treatment. However, the operation on the beating heart is complicated for a surgeon, technically challenging, and time consuming.

The reason is that without immobilization, the local myocardial stabilization by current vacuum-assisted stabilizers cannot completely eliminate the heart motion [180]. According to the medical study [95], they may reduce the motion amplitude by only about 50 %. Since the remaining complex motion of the beating heart exceeds the human tracking bandwidth of about 1 Hz, the manual synchronization of surgical instruments with the heart motion is very challenging for a surgeon [66]. Furtheremore, the high accuracy is necessary for ensuring safe surgical interventions. As reported in [66], similarly to the surgical operations on a rested heart, the tracking precision must achieve from 0.1 mm up to 0.2 mm depending on the diameter of coronary arteries ranging from smaller than 1 mm up to 2 mm. Moreover, the performance of the beating heart operation is additionally impeded by a tremendous amount of hand-eye coordination accompanied by short reaction time to rapid changes in heart behavior, e.g., by extrasystoles.

Introduced in beating heart operations, the computer-assisted surgery system aims at enabling surgeons to overcome these limitations. It allows for an increase tracking accuracy, improve dexterity, and hand-eye coordination.

Computer-Assisted Surgery System This system, which implements the task of the beating heart motion compensation, is of particular importance for beating heart surgery. As schematically illustrated in Fig. 1.1, it addresses the problem of synchronizing surgical instruments with the beating heart so that the relative motion between the instruments and the heart is minimized. For this purpose, the heart is observed by cameras installed above an operation table for open thorax surgery [19, 178] or integrated in an endoscope for minimally invasive surgery [150, 181, 197]. Additionally or even alternatively, the measurement information can be provided by medical imaging modalities, e.g., ultrasound [84, 229, 230], electrocardiogram [23, 160], a respiration pressure signal [159, 160], or other sensor systems like acceleration sensors [82, 144], sonomicrometry [23, 45, 46], force sensors [29, 40, 230], or whisker sensors [24]. By utilizing the estimated position of the heart, a robot synchronizes surgical instruments with the heart. At the same time, it moves the surgical instruments to the position predetermined by a surgeon.

Another important aspect of this system is that the surgeon gets an impression of operating on a non-beating heart and can therefore precisely define surgical interventions. The surgeon navigates the robot for performing

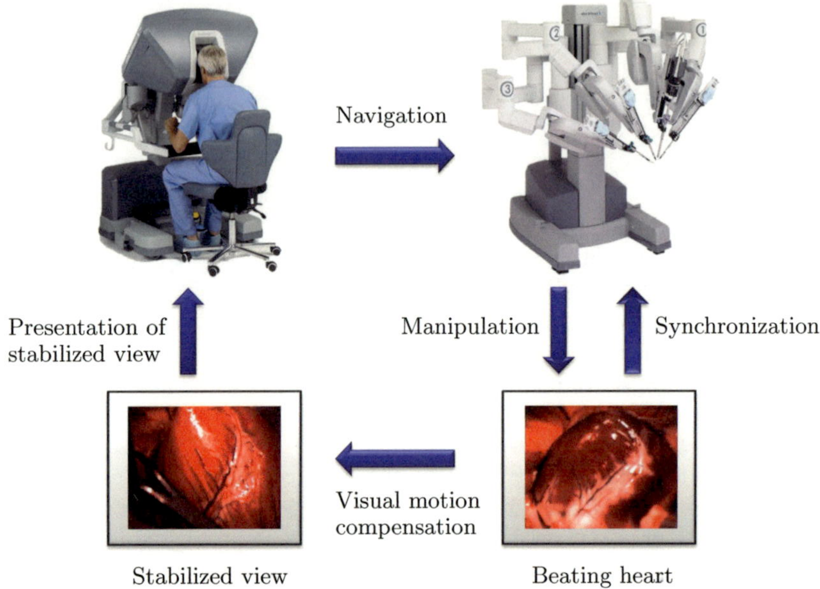

Figure 1.1: A computer-assisted surgery system assists a surgeon at operations on a beating heart by synchronizing the surgical instruments with the heart surface motion. The beating heart is represented to the surgeon as virtually stabilized for precise definition of the surgical interventions (source of upper images: http://www.intuitivesurgical.com/).

surgical manipulations, such as cutting or burning of heart tissues, based on a virtually stabilized view provided by a visual motion compensation.

Taken as a whole, the beating heart surgery system is utterly dependent on the heart motion tracking and reconstruction quality. Significant for synchronization of surgical instruments, robotic navigation, and manipulation, as well as visual motion compensation, motion tracking is one of the essential components of the surgery system.

1.2   Problems under Consideration

The main challenges for this work are:

- Prediction and tracking of the three-dimensional position of the entire elastically deformable object.

- Visual motion compensation by stabilizing a view of the object, wherein the goal is to represent this object as non-moving.

Both problems cover **four important aspects** associated with reconstructing the three-dimensional motion of the spatially and temporally varying elastic object from noisy space- and time-discrete measurement data.

1) Noisy measurements The elastically deformable object such as, e.g., the beating heart, is supposed to be observed by a stationary camera system. Commonly, the acquired measurement information is not reliable enough with regard to the specification of the exact position of the object under observation even on measurement points. The reason for this is the fact that the measurements extracted from camera images are contaminated by noise caused by the inaccuracies of the image formation and processing.

2) Reconstruction of entire object The measurement information about the object's deformation is limited to some discrete points at the object's surface. The reason for this lies in the fact that the camera images are represented by a discrete set of pixels with assigned colors. Therefore, the second aspect throws up the challenge of reconstructing the motion of the entire object, i.e., also between measurements points. It should be pointed out that the motion of the object can significantly differ from one point to the other because of elastic deformation.

3) Loss of measurements The accuracy of the motion compensation strongly depends on the amount of measurement points. Unfortunately, most practical applications that use cameras suffer from loss of measurements due to the sensitivity of a camera to illuminations, dust, dirt, and occlusions. For instance, a beating heart operation is unimaginable without concealing any part of the heart, e.g., by surgical instruments or escaping blood in the field of camera view. All of the foregoing can lead to the partial and even total loss of measurements over a long time horizon.

4) Unknown physical properties of object Finally, the problem of tracking and motion compensation is exacerbated by unknown physical properties of the object under observation, such as elasticity or density. These properties are of particular importance when no measurement information is available and therefore, only a priori knowledge can be used for estimating and predicting the position of this object. Moreover, the object's behavior due to changing environmental conditions like external excitation is also unknown.

1.3 State of the Art

This section provides a survey of methods relating to the motion compensation of elastically deformable objects. Since the motion compensation is as accurate as the reconstruction of the entire object's motion, the methods for tracking and visual motion compensation are classified according to two aspects that contribute to the accuracy of the motion reconstruction. The first aspect concerns the models embedded in the processing scheme. The second aspect relates to the types of uncertainties considered by motion reconstruction.

1.3.1 Model- and Non-Model-Based Methods

According to embedded models, the method for tracking and visual motion compensation are primarily divided into two groups: model-based and non-model-based. The aim of this section is to give an overview of the models on which these methods are based. In contrast to recent surveys [83, 132], the analysis of the existing models is mainly focused on the beating heart surgery application, while providing an insight in the methods applied in other fields where necessary.

Non-Model-Based Methods

The non-model-based methods process measurement data without explicit models of the object under consideration and measurement process, and therefore incorporate little or even no a priori information in the motion reconstruction.

On the one hand, these methods are very sensitive to the measurement noise and outliers since they reconstruct the object's motion based only on the measurement information. Furthermore, they fail in case of complete loss of measurement information. On the other hand, without considering a priori information, the non-model-based methods are independent from the object under observation and therefore are of low computational complexity, as illustrated in Fig. 1.2.

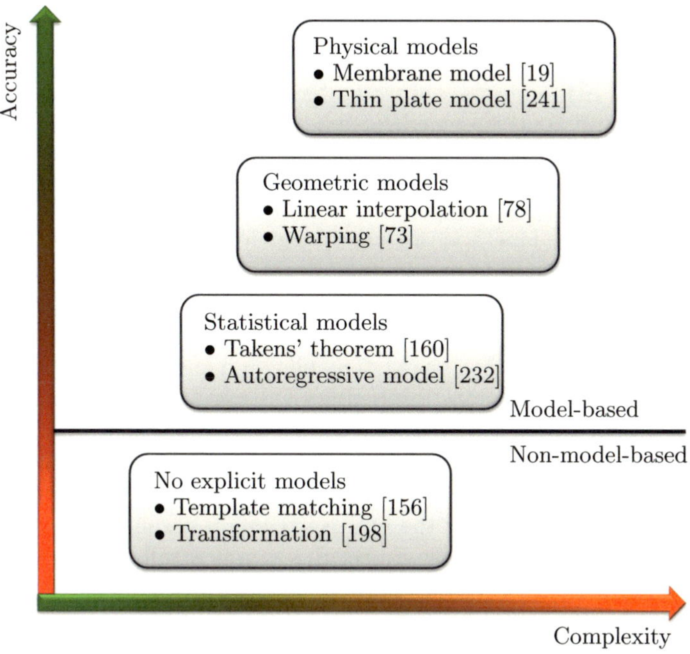

Figure 1.2: Overview of possible methods for motion compensation according to the models used for processing measurement data. The incorporation of a priori information in processing of measurement data increases the accuracy of the motion compensation but simultaneously leads to higher computational complexity.

Generally, the non-model-based tracking methods are widespread. For the purpose of providing some examples, the tracking methods based on finding similarities between patterns in consequent image frames [92, 100] or optical flow [204] refer to this group. In the context of the motion **tracking** for beating heart operations, the methods proposed in [23,156,170,181,197] can be assigned to this category. For instance, in [181], the motion of the heart is reconstructed on some points using pattern-based tracking, which involves an efficient second order method proposed by [28]. The idea of this method is to determine the homography, which minimizes the sum of square differences between the reference and current patterns. In [156], the template matching and texture analysis build a basis of the composite tracking algorithm. Moreover, the proximity and similarity measures introduced in [164] and image warping based on the Lucas-Kanade registration algorithm [133] extended for incorporating stereo constraints are used in [197] for reconstructing the motion of the heart surface from image sequences provided by a stereoscopic laparoscope.

The non-model-based methods for **visual motion compensation** are primarily hardware-specific. For example, in [71] an electrocardiogram triggered strobe light is applied for the purpose of making the heart appear still to a surgeon. Instead of improving the quality of the beating heart operation, the authors report increasing demands on a surgeon's concentration and fatigue due to deliberate tampering of the image sequence. According to [150], the stabilized view is created by synchronizing the camera with the heart surface motion. Another kind of camera synchronization is proposed in [198], wherein the current camera image is transformed in a stabilized image by changing the extrinsic camera parameters. It should be noted that all these methods expect that a very small and smooth area of the heart surface is observed. They compensate only the global motion of the heart surface, whereas local motion distortions, which are defined by a difference in the motion of the neighboring points, cannot be eliminated.

Model-Based Methods

The model-based methods distinguish themselves by exploiting different models for processing the measurement data. In contrast to non-model-based methods, the model-based methods enable a priori information to be involved into the problem of extracting the desired information from noisy measurements. Furthermore, they allow for bridging the loss of measurements by means of reconstructing the object's behavior based on a priori information. This certainly is beneficial for increasing accuracy and robustness of the motion compensation.

The essential problem of model-based methods lies here in the dependence of the computational complexity of these methods upon the reliability of the models used. For example, a high accuracy of the methods for tracking and visual motion compensation is accompanied by a high burden of computational complexity, as illustrated in Fig. 1.2. Continuing this line of thought, a highly accurate motion reconstruction can be achieved by physics-based methods incorporating a high amount of a priori information about the physics of the object under observation. Unfortunately, these methods governed by partial differential equations are most computationally expensive, hindering their application in real-time systems.

With the intention to give an overview of existing model-based methods, the methods for tracking and visual motion compensation are classified according to the models used.

Statistical Models The methods that are based on statistics for analyzing time series data [42], i.e., Fourier series, or autoregressive models, refer to the statistical category. By considering only limited physical information characterizing the system response, e.g., resonance frequency in Fourier series models, these methods can lead to incorrect motion prediction and reconstruction, especially in case of strong measurement artifacts or rapidly changing behavior of the object.

When **tracking** methods for beating heart surgery are considered, the method proposed in [160] is associated with this group. Here, in case of occlusions, the beating heart motion is reconstructed using the correlation with the measurements lying in the past and Takens' theorem. Furthermore, in [231], a Fourier series model is used for reconstructing the motion of the mitral valve by processing ultrasound measurements. In further publication of this author [232], an autoregressive model is proposed.

Commonly, the methods using time series models consider only some points on the surface of the object under consideration. In particular, the spatial dependencies, such as motion between these points, are ignored. Hence, reconstructing the motion of the entire object will overwhelm the model dimensions as every point of the object must be included in the model. Perhaps, this is the main reason that these models are not applied for **visual motion compensation**.

Geometric Models The methods that incorporate models of an object geometry, such as approximation of the object surface by B-splines or radial basic functions [235], geometric warping [73], or morphing [226], refer to the geometric category. In this context, the spatial dependencies between the object points are incorporated in the motion reconstruction and determine the predicted behavior of the object. However, the approximation of complex object's behavior may be incorrect, as no physical properties of the object are incorporated.

However, geometric models are widely used for **tracking** of elastic objects [21, 114, 161, 175] because of their simplicity and computational efficiency. In case of high camera frame rate and reliable measurement information, they may provide satisfactory results. Further advantage of these models is a small number of parameters needed to describe an object. For example, in [21] a parametric deformable model is proposed for tracking of the deformation of the left ventricle in a single-photon emission computed tomography. This model is represented by a free-form shape that is build upon a superquadric fit. In order to track some points on myocardium in magnetic resonance imaging, [161] exploits parameterized volumetric primitives, which deformation is defined by few locally varying parameter functions. In [49], the left ventricle is modeled as a tapered ellipsoid with parameters determined by an optimization schema that fits the ellipsoid's shape to available data. The tracking method proposed in [175] aims an application in beating heart surgery system. Here, the heart surface motion is determined by constructing the minimization problem involving thin plate spline functions. In [114], the authors reconstruct the heart surface by optimizing a parametric B-spline function.

As for **visual motion compensation**, the image of the beating heart is stabilized in [78] by a linear interpolation. This method is based on the transformation of the images acquired at different time steps to the chosen reference image by exploiting a triangulation of the object geometry. Unfortunately, this method requires a fine discretization of the object geometry for achieving a high quality of the stabilization. Furthermore, the stabilization may become rough [73]. Another example of geometric methods are region-based deformable appearance models presented in [186]. They incorporate a combined parametrization provided by modal analysis and principal component analysis. Although the modal analysis is originally physics-based, their non-physical parametrization destroys the physical meaning. A further classical example of visual motion compensation consists of exploiting geometric primitives, e.g., ellipsoids, as proposed in [5], where the stabilized image sequence is created for frontal face recognition.

Physical Models   In contrast to the above-mentioned methods, the physics-based methods for motion compensation incorporate physical models that operate directly in the physical space. These methods involve not only geometrical but also physical a priori information, which is defined by the material structure of the object under consideration. When the embedded physical model is appropriate, the physics-based methods provide a physically correct and therefore, in comparison to other methods, more accurate reconstruction of the object's motion by processing the noisy measurement data. The major problem of the physics-based methods lies particularly in the fact that these methods are most complex and demand much computational power. The main reason for this is that highly detailed models of complex geometry and material structure are required for accurate approximation of the complex behavior of the real object. There are different types of physical models that can be found in the related work.

Mass-Spring-Damper Models To the best of our knowledge, a mass-spring-damper modeling technique is introduced in [165] for a facial animation and is first applied for surgical simulations in [56]. In the latter work, a simple surface mass-spring model for simulating deformations of a gall bladder is proposed. The key idea is to represent the object by a set of mass points connected by springs and dampers. The motion of every mass point is defined by an equilibrium of the inertial force determined by Newton's second law and forces acting on each point from the springs and dampers.

One of the difficulties of the mass-spring-damper models is to identify the stiffness of various springs reflecting the physical properties of the object under observation. With respect to this problem, it is evident in [70] that the modeling of homogeneous materials is generally not possible. The other problem is the large number of model parameters needed for realistic representation of material behavior. The reason for this is a large number of springs, which usually connect each mass point not only with neighboring points but also with many other mass points. These connections are necessary for the efficient transfer of forces. Furthermore, the modeling of incompressibility or transverse contraction is problematic without additional penalty forces [81]. However, in spite of poor precision and stability problems [34], these models are favorable in computer graphics and surgical simulations because of their simplicity, ease of implementation, and reasonable execution time.

Particle Models The concept of particle models, pioneered by [145] for handling problems in compressible gas flow, is more general than that of mass-spring-damper models. These models preserve an idea of discretizing an elastic object by a set of mass points. However, these points are called particles because they possess individual material properties, such as, for example, density or stiffness. The propagation of particles through space and time is defined by partial differential equations, derived from the physical principles underlying the motion of the object under observation. It should be noted that in contrast to mass-spring-damper models, the interactions between the particles are not restricted to damping and stiffness forces but can be defined by arbitrarily shaped forces. The advantage is that these models do not use a fixed topology as mass-spring-damper models do. This is why they are commonly applied for modeling physical phenomena without a fixed neighborhood, such as water or fire [3,54]. Furthermore, since the connectivity between the nodes is computed at every time step and therefore, can easily be changed with time, these models are especially suitable for handling large deformations of objects [60,124].

However, with respect to modeling of elastically deformable objects, one of the important disadvantages of such models is the difficulty to represent a smooth surface with particles. This is due to the fact that particle models do not handle the surface itself, and thus demand a robust computation of its curvature that can be cumbersome [54]. This is also the reason that handling of boundary conditions is recognized as a difficult task has been tackled by ghost particle method [128], distance functions [79], and direct forcing method [26].

Continuum Models In contrast to particle models, these models are based on solid mathematical and physical foundation, e.g., introduced in [33]. By representing an object as a continuum, they describe its deformation at every point of space at every time. This implies that the changes of the object are assumed to be continuous, so that there are no changes in the small neighborhood of a point in the undeformed and deformed state [11]. The motion of the continuum is governed by partial differential equations constructed based on physical principles, such as the principle of energy conservation or the principle of virtual work.

The drawback of these methods is their high computational cost leading to trade off accuracy for real-time functionality. Often, the partial differential equations are not analytically solvable. Therefore, computationally expensive numerical methods are needed for approximating their solution, e.g., finite element method or finite difference method. Moreover, solving these equations numerically leads to handling of high-dimensional and often nonlinear systems.

However, the continuum models are well-established in different applications, e.g., in computer graphics modeling [96, 193, 209], facial image analysis [119, 210], medical imaging [58, 139, 140], or preoperative planning and diagnostics [188]. Commonly, in these fields, real-time functionality is not the most important aspect and approximation of the object's behavior in a most realistic manner stands in the foreground. For example, the heart behavior is described in [188] by a highly detailed finite element model with complex geometry and material characteristics.

Survey of Physics-Based Methods for Motion Compensation   Unfortunately, due to stringent requirements for real-time functionality, the application of physical models for a beating heart robotic surgery system is hardly possible.

With the intent to meet these requirements, some **tracking** methods were proposed. For example, the more efficient but less accurate finite element model is introduced in [177] for tracking of the landmarks on the heart surface. However, although material characteristics of the heart are strongly simplified, this model suffers from an unnecessarily complicated description of the heart geometry. Even stronger assumptions are introduced in [19] in order to enable real-time motion tracking. This method approximates the heart surface with a membrane model, which is utilized by a Kalman filter estimation. It should be noted that the motion of the landmarks is reconstructed here in only one direction. Furthermore, the initial configuration of the model is assumed to be known and very simple constant boundary conditions of Dirichlet type are defined. Another physics-based method for tracking the heart motion is proposed in our paper [241], wherein the heart wall is modeled as a thin plate. Here, it is assumed that the out-of-plane deflections of the heart surface are small and that the thickness of the heart wall remains constant during the heart deformation. This allows drastically decreased computational complexity by reducing the three-dimensional model of the heart wall to a two-dimensional one.

However, all these methods share the same limitation: The physical models are restricted to objects with known and constant physical parameters. Without adjusting the model parameters to changing behavior of the objects under observation, these methods cannot compensate for the differences between the individual objects. Another disadvantage is that the most of them apply a predefined mesh, e.g., finite elements in [19, 177], for discretizing the physical model. On one hand, this can lead to high numerical errors if the mesh is too coarse for reconstructing the large deformations of the object. On the other hand, models can be uselessly complex if the mesh is unnecessarily fine.

In respect to the methods for **visual motion compensation**, it should be noted that early attempts at introducing the physics-based deformable models for processing of image sequences are made in [162] with the aim of animating realistic facial expressions from images. Furthermore, in [210], deformable contour models, also called snakes [105], are combined with the three-dimensional physical model for estimating the face muscle contractions from the image sequence. Although the physical models are further explored for facial animations [64,120,234] and have been already proposed for video coding [86], they are still rare in image processing due to their enormous computational complexity caused by the demand for a realistic representation of the physical objects with complex dynamics. Accordingly, to the best of our knowledge, there are no physics-based methods for visual motion compensation.

1.3.2 Deterministic and Probabilistic Methods

One of the reasons for the deterioration in accuracy of the motion compensation is that the measurements are corrupted by disturbances. As illustrated in the previous section, the models involving a priori knowledge may be utilized for precisely extracting the desired information from measurement data, which are susceptible to errors. Unfortunately, these models are only inaccurate representations of the real objects and also introduce errors in the motion compensation. Before reviewing the existing methods according to their handling of uncertainties, this section classifies the uncertainties.

Types of Uncertainties Generally, there are two different types of uncertainties: stochastic and systematic. The stochastic uncertainties characterize random errors. These errors can be described statistically and reduced by averaging the repetitive measurements. For example, the stochastic uncertainties of the camera measurements stem from instrumental random errors, such as electronic noise, flickering, imprecisions of feature extraction or shape distortion [51]. In the motion models used for measurement data processing, these uncertainties can be inherited from input noise, as well as from uncertainties of parameters. In contrast to stochastic errors, the systematic uncertainties are difficult to detect, especially if they change over time. Commonly, these errors are associated with calibration inaccuracies of the measurement systems and limitations of the underlying models. For example, the systematic uncertainties of camera measurements can be caused by instrumental systematic error [51], which is associated with the restricted validity of the calibration model, e.g., due to incorrect estimation of model parameters or nonlinearities that this model cannot take care of.

With respect to handling of uncertainties, all methods can be divided in three groups, as illustrated in the Fig. 1.3.

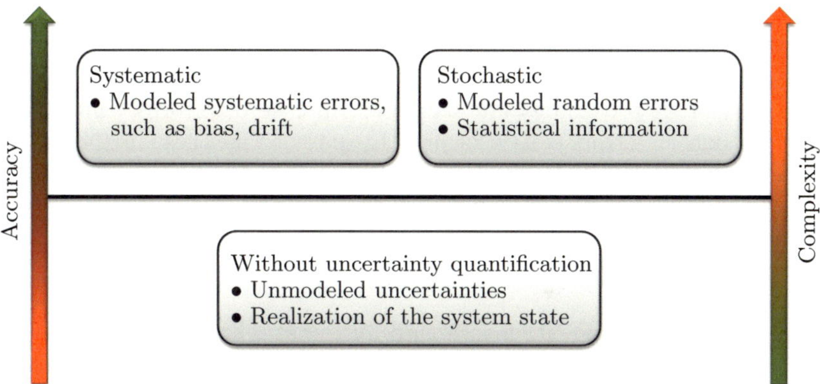

Figure 1.3: Overview of the possible methods for motion compensation according to types of uncertainties considered. Consideration of stochastic and systematic uncertainties incorporates more a priori information in motion compensation. The main challenge of motion compensation is to combine high accuracy of the object's motion estimation with low computational complexity.

Methods without Uncertainty Quantification   The methods that do not consider the uncertainties at all [57,160,182] refer to this group. By assuming that the measurements are deterministic, i.e., without random variations, these methods provide one possible realization of the random object deformation. This assumption can only be maintained if measurement disturbances are negligible. When the measurements as well as the models utilized are affected by high errors, incorrect results can be obtained.

This is especially dangerous for the safety of the beating heart operations, where wrong behavior of the surgical robot can lead to perilous injuries.

Methods Considering Stochastic Errors   An increasing number of methods for estimating heart surface motion cope with stochastic uncertainties [19, 174, 177, 231, 232]. By applying Bayesian inference scheme, these methods commonly provide statistical information about how accurate the motion estimate is. In addition to to the system model, the measurement model, as well as models of accompanied noise, are incorporated in the motion estimation. For linear systems contaminated by Gaussian noise, a Kalman filter [101] is employed for obtaining an optimal estimate. For non-linear systems, nonlinear estimation procedures, such as extended Kalman filter [222], are used. However, due to the assumption that the system state is characterized by a unique probability density, a Bayesian inference scheme provides correct estimates only when the models underlying the estimation procedures are accurate. In case of incorrect modeling, high estimation errors can occur [90]. In spite of this, these methods lack a continuous evaluation of the quality of the models and precise analysis of the modeling and measurement uncertainties. However, neglecting these errors leads to a lack of precision that increases greatly the risk of complications during beating heart robotic surgery.

Methods Considering Stochastic and Systematic Errors   Consideration of both stochastic and systematic errors is computationally more expensive but at the same time more accurate than accounting for only stochastic errors. Naturally, the reason for a better accuracy is a higher amount of a priori information incorporated in the motion estimation. Although accurate motion estimation is crucial for the safety of beating heart surgery, to the best of our knowledge, these errors are still neglected in existing methods for heart motion compensation.

1.4   Novelties and Contributions

This thesis proposes an efficient and accurate framework for motion compensation of elastically deformable three-dimensional objects. The core of this framework is a physical model, which serves as a starting point for the derivation of all other mathematical models. As the physical models are object-specific, the motion of the beating heart with regard to the application in a beating heart surgery is in the focus of this work.

In this thesis, to our best knowledge, the first **volumetric physical model** of the heart wall suitable for this application [240] is proposed. Describing surface and interior of the heart wall, this model reflects the volumetric behavior of the heart, i.e., motion of the heart surface is influenced not only by surface points but also by the points inside of the organ. In this way, a foundation for modeling surgical procedures during the beating heart operations is built.

The main advantage of the proposed model is the combination of **high precision with reconcilable complexity**. This is achieved, first of all, by focusing on the part of the heart ventricle that is essential for beating heart operations. Due to small displacements of this part, its behavior is mathematically formulated by a system of linear stochastic partial differential equations[1] with random input and uncertain initial and boundary conditions. Instead of numerical solution of this system using classical finite element method, an efficient element-free method is employed that discretizes the spatial domain of the model by a set of points, without introducing a predefined connection between them.

One of the major characteristics of the framework is in a simplified mathematical description of the heart behavior combined with a detailed description of modeling and measurement errors. An efficient handling of **stochastic and systematic errors** in the context of beating heart operations was first introduced in our works [238] and [237], to the best of our knowledge. The consideration of these errors is crucial not only for reduction of the computational complexity but also for achieving high accuracy of the reconstruction. Therefore, the physics-based models, such as system model and measurement model derived from the system of stochastic partial differential equations, incorporate terms that model stochastic and systematic errors. In this thesis, the framework copes with systematic errors by their augmentation with the system state and simultaneous state and parameter estimation. The quantification of stochastic errors is incorporated in this estimation.

As far as we know, existing methods for visual motion compensation do not consider physical properties of the object under observation. Here, a

---

[1]The system of stochastic partial differential equations results from a system of partial differential equations, where the coefficients, input, initial, and boundary conditions are considered as random variables. The solution of this system is a random field consisting of the infinite set of space- and time-dependent random variables.

novel physics-based **method for visual motion compensation** is introduced. It exploits the image transformation function with physically justified parameters [237]. Additionally, an **adaptation method** for continuous monitoring and improvement of the motion compensation quality [236] using a feedback mechanism is proposed. This method allows to reduce discretization errors in areas where motion compensation is inaccurate: The model discretization is refined by inserting additional points with assigned physical parameters. Furthermore, the physical properties of the model are flexibly adapted with the aim to adjust the model to the behavior of the object. This is achieved by the consequent adjustment of all model parameters using a simultaneous state and parameter estimation [239].

With respect to the beating heart motion compensation, the proposed methods offer a broad range of benefits including physically correct motion prediction and estimation of the entire interventional area of the heart. The reason is that due to incorporating the physical model of the heart wall, a high amount of a priori knowledge about the geometrical and physical characteristics of the heart is involved in the motion reconstruction. This greatly enhances the resolution of motion compensation. Furthermore, it is the main reason that the framework is robust to partial and total loss of measurement information, e.g., caused by occlusion of camera views. A reconstruction of the occluded part of the heart is especially essential when the beating heart is concealed by surgical instruments or robotic arm during the operation. Moreover, the methods are expected to be flexibly adjustable to different patients due to continuous evaluation of the motion compensation quality and its online adaptation.

1.5 Outline

The thesis outline is illustrated in Fig. 1.4, where the direction of information flow from one chapter to the other is indicated by arrows. In the following, the content of every chapter is briefly introduced.

Chapter 2 points out main ideas of tracking and visual motion compensation methods. This includes a physics-based interpretation of the entire framework, a systematic derivation of all mathematical models from an underlying physical model of the object under observation and, finally, a flexible adjustment of accuracy and complexity of the models using stochastic estimation and feedback mechanism.

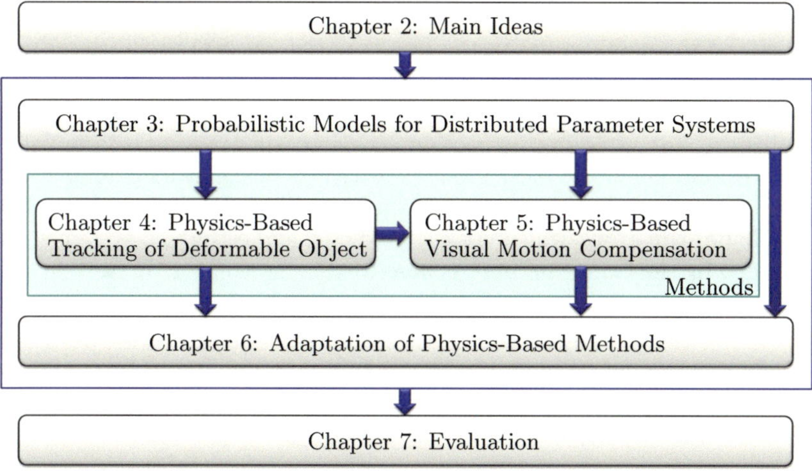

Figure 1.4: Thesis outline with interconnections between the chapters. The probabilistic models build a basis for predictive tracking and visual motion compensation methods. By the adaptation of all mathematical models, the quality of the entire system is improved.

Chapter 3 is concerned with the derivation of the models for estimation of the heart wall motion. For this purpose, first of all, a novel physical model approximating the heart wall behavior is established. It builds the basis of all proposed methods. First, the mathematical formulation of this model, in form of a system of stochastic partial differential equations, is deduced from physical principles underlying the behavior of the heart wall. Then, in order to enable the stochastic estimation of the object's position at every time step, the physical model is converted in a discrete state-space form, consisting of the system model, given in Section 3.3, and the measurement model, proposed in Section 3.4. Here again the modeling and measurement errors are described and their evaluation is modeled. For conversion of the model into the space-discrete form, an efficient element-free numerical method is applied. In this context, an overview of numerical methods for spatial discretization of the stochastic partial differential equations is given in Section 3.3.1.

Chapter 4 presents the predictive tracking approach. The main characteristic of this method lies in the fact that the physical properties of the object under observation are involved in the motion prediction and estimation. The probabilistic problem that this method solves, as well as its challenges, are introduced in Section 4.1. Here, the tracking is formulated as an estimation of the most probable positions of the heart landmarks at every time step. The estimation of these positions by means of physics-based approximation, which use the predicted and estimated system state provided by the Gaussian filter, is described in Section 4.4. In Section 4.3, the predicted positions of the landmarks are exploited for extraction of the measurement information from the camera images. For this purpose, the correspondence function is defined, which, in addition to the standard criteria, incorporates the physics-based criterion for establishing unique correspondences between the landmarks and measurements.

Chapter 5 deals with a novel method for visual motion compensation. First, the problem of visual motion compensation is formulated as a transformation of image sequences in Section 5.1. Then, the key idea of the proposed method, i.e., the three-dimensional physics-based image transformation, is presented in Section 5.2. Thereafter, the physics-based image transformation function is established in Section 5.3 by projecting the current position of the object under observation onto the image plane of the camera. Finally, a stabilized image sequence is obtained in Section 5.4 by transforming every image of the image sequence provided by a camera to the reference image.

Chapter 6 introduces an adaption of the underlying physical model to the behavior of the object. All required models for tracking and visual stabilization are refined with respect to their spatial resolution. Furthermore, they are extended for considering the inhomogeneity of the material of the object. The adaptation strategy is introduced in Section 6.1. The specialty of this strategy is to use a feedback from the visual motion compensation for adaptive improvement of the quality of the system only where necessary to increase the resolution of the models. The extraction of the feedback information from stabilized image sequence provided by the visual motion compensation is described in Section 6.2. Consequently, in Section 6.3, the discretization of the physical model, and all derived from it physics-based models, is refined with the aim to increase their spatial resolution and reduce parameter errors.

Chapter 7 presents the evaluation of the proposed methods in respect to their application in a beating heart surgery system. The quality of the methods is verified in the experiments on the pressure regulated artificial beating heart, which simulates the motion of the mechanically stabilized real heart. After introduction of the experimental environment and error measures used for quantification of the quality of the system in Section 7.1, the experimental results are presented in Section 7.2. Here, the system for motion compensation is tested regarding its sensitivity to the loss of measurement information and its capability to compensate the changes of the heart motion frequency.

Chapter 8 closes the thesis with the discussion of the main points of the proposed methods, the remaining challenges and future work.

1.6 Summary

In this thesis, physics-based framework for motion compensation is proposed. It is mainly dedicated to beating heart surgery. For this application, the tracking of the three-dimensional heart motion is essential for synchronizing surgical instruments, robotic manipulation and navigation. The visual motion compensation aims at giving a surgeon an impression of operating on a non-beating heart by presenting him a virtually stabilized heart view.

To the best of our knowledge, this thesis is the first to propose volumetric physical model of the heart wall for the application in beating heart robotic surgery system as well as physics-based method for visual motion compensation. As all mathematical models incorporated in the framework are derived from the heart wall model, the physical principles describing the motion of the heart wall are directly incorporated in the motion compensation. This ensures a physically correct motion reconstruction that is especially crucial when the measurement information is lost.

Another important characteristic of the framework consists in the combination of the simplified mathematical description of the heart behavior with a detailed description of the modeling and measurement errors. In this way, in contrast to existing physics-based methods, one of the main challenges of the physics-based formulation has been met, i.e., achieving high accuracy and computationally tractable complexity. Furthermore, a consequent monitoring and improvement of the tracking and visual motion compensation quality is ensured by a feedback mechanism. This enhances the spatial resolution of the physical model and therefore, the spatial resolution of all mathematical models incorporated in the framework.

2 Main Ideas

This chapter describes the key ideas of the proposed methods for motion compensation:

1. Physics-based formulation of the tracking and visual motion compensation that allows to introduce physical knowledge into the motion reconstruction of the object under observation.

2. Systematic derivation of all mathematical models, exploited by both methods, from one physical model of the object under observation.

3. Flexible adjustment of accuracy and complexity of both methods using a stochastic estimation and a feedback mechanism.

2.1 Physics-Based Formulation of the Methods

In contrast to standard methods that are non-physics-based, the methods proposed in this thesis are based on the fact that the measurement data provide information about the motion of the physical object under observation. Therefore, these data are influenced by physical characteristics of the object.

Within this context, the tracking method aims at determining the three-dimensional positions of a number of surface points at every time step, based on images provided by a stationary camera system, as shown in Fig. 2.1. It is straightforward that the displacements of these points are influenced by physical characteristics of the object, e.g., material density or elasticity.

As for the visual motion compensation, its aim is to represent the dynamic scene under observation as stationary. This is accomplished by an image transformation, which transfers the colors[2] of the current image acquired by a camera at time step t_k to the appropriate positions in the reference image, e.g., acquired at time step t_{k-n}.

---

[2]For describing the colors, different color models may be applied. For example, according to RGB color model, which stands for red, green, and blue, the colors can be obtained by combining these three lights, also called color components, in varying intensities.

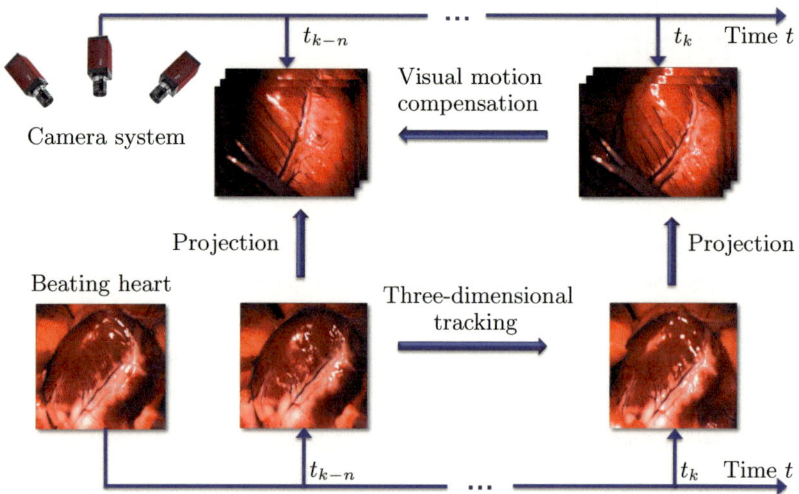

Figure 2.1: Key idea of physics-based formulation of the tracking and visual motion compensation. Each image can be interpreted as a projection of the object under observation onto the image plane of the camera. This allows the introduction of the physical knowledge about the object into the reconstruction of the object's motion and each camera image.

The physical information incorporated in the camera image is established by the fact that the image may be interpreted as a projection of the physical object under observation onto the image plane. This means that the current position of the physical object determines the positions of the pixels in the current image. Therefore, when the position of the object at two different time steps, e.g., t_{k-n} and t_k, is known, the correspondences between the pixels, as well as their positions in the appropriate images, are determined. As the position of the object under observation is bounded by physical characteristics of the object, the pixel positions are bounded as well.

It should be noted that the physical interpretation builds the basis of both methods. This ensures the physically correct image reconstruction along with the reconstruction of the object's position.

2.2 Systematic Derivation of Mathematical Models

All mathematical models are derived from a physical model, which approximates the behavior of the physical object under observation, as shown in Fig. 2.2.

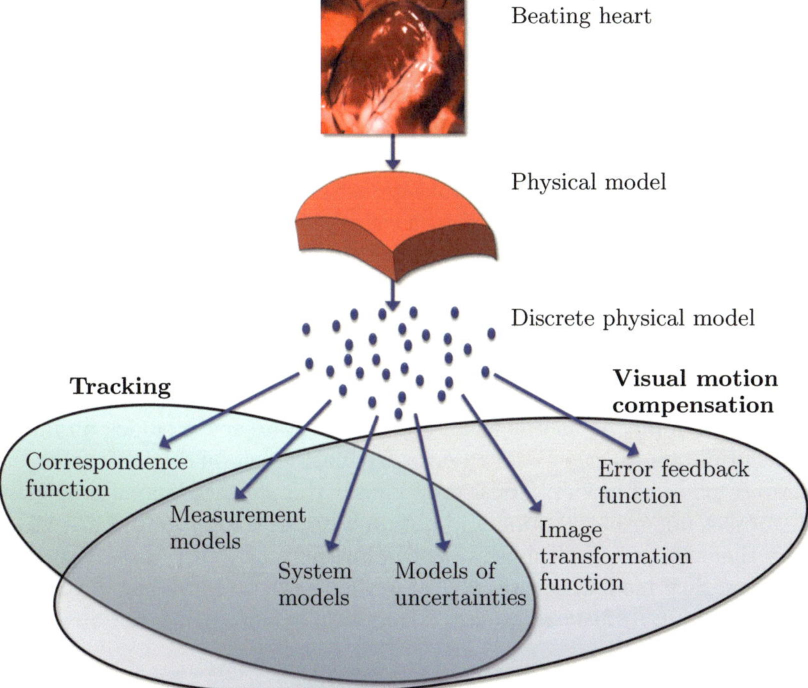

Figure 2.2: All models exploited by the tracking and visual motion compensation are derived from one physical model, which approximates the behavior of the object under observation.

Such a systematic derivation differentiates the proposed methods from many others that are based on diverse heuristics.

The key idea of this approximation is to get insight into the complex behavior of the real object. By approximating the object directly in the physical space, the model incorporates physical characteristics of the object, such as, e.g., elasticity or damping. The reason for this approximation is the unknown behavior of the real object, so that exact positions of the points on the surface of the object are unknown. Although the positions of some points can be measured, the measurement data obtained are corrupted by uncertainties due to, e.g., inaccurate camera calibration or illuminations. By simplifying the object's behavior, the model allows for reconstructing the positions of all surface points at every time step based upon measurement information extracted from camera images.

However, the main problem here is that the motion state of the physical model cannot be obtained analytically from a system of stochastic partial differential equations, which describes the motion of the physical model. In order to make the numerical computations tractable, the spatial domain of the physical model is discretized by a set of points. It should be noted that this discretization is called element-free since no predefined information about the connection between the discretization points is necessary.

All mathematical models incorporated in tracking and visual motion compensation methods are derived from the discrete physical model. As illustrated in Fig. 2.2, a correspondence function, a measurement model, and a system model, as well as an augmented state-space model, an image transformation function, and an error feedback function originate from the discrete physical model. Thus, they inherit the physical characteristics of the object under observation and obtain a physical interpretation that allows them to be called physics-based. Furthermore, such a derivation of mathematical models ensures a clear system structure and has the advantage of simultaneously improving all models by increasing the quality of the physical model.

2.3 Flexible Adaptation

One of the main characteristics of the proposed methods is the flexible adjustment of their accuracy and complexity. It is achieved by systematical improvement of the models involved in the methods. This is based on the feedback from visual motion compensation as well as simultaneous state and parameter estimation, as schematically illustrated in Fig. 2.3.

Naturally, the quality of the model-based methods strongly depends upon the quality of the models. Although the model is only an inaccurate approximation of the object's behavior, most existing methods [64, 120, 186,

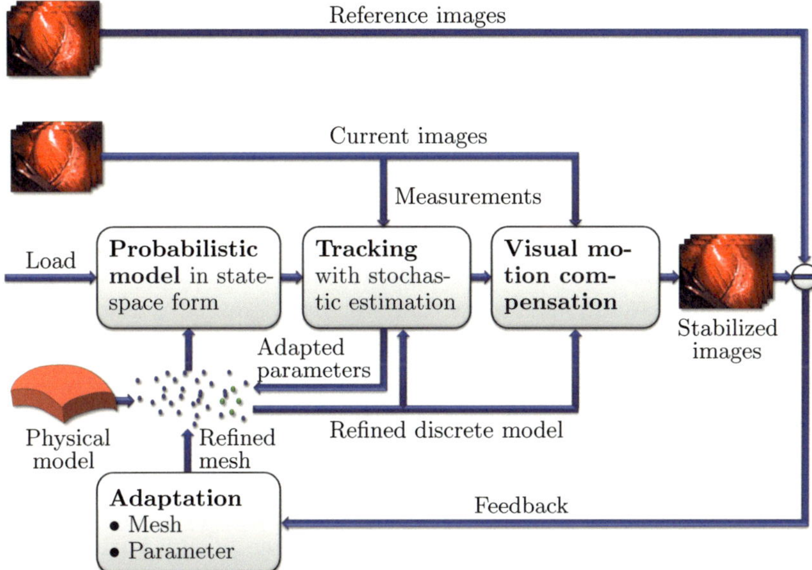

Figure 2.3: Adaptation of the discrete physical model leads to the improvement of the entire system. The predictive tracking approach determines the parameters of the model by processing the incoming camera measurements. The feedback from visual motion stabilization is used for adaptive refinement of the discrete physical model, whereby the spatial resolution of this model is improved only where necessary according to the feedback information.

234] assume that the model describes this behavior exactly. They neglect to take into account stochastic and systematic errors, such as, e.g., noise of the model input, incorrect model parameters, bias, or numerical errors, which lead to a deviation of the model behavior from the behavior of the object under observation. However, the model-based methods provide a superior accuracy only when the incorporated models approximate the object's behavior accurately.

In this thesis, the divergence between the physical model and the object under observation is sustainably reduced by an adaptation according to the scheme illustrated in Fig. 2.3.

Accordingly, the material properties of the model as well as its discretization are refined. It should be noted that the material properties of the model are initially defined by a minimal number of parameters, which describe the heart wall as a homogeneous viscoelastic body. Therefore, when the material of the object is inhomogeneous, high errors may arise in motion compensation. Furthermore, the model is discretized by a set of points, also called model nodes, depicted in Fig. 2.3 by blue points. When the spatial discretization of the model is coarse, errors may arise due to low spatial resolution. For example, when the motion of the object varies strongly from point to point, a fine discretization is necessary for reconstructing the object's displacement precisely.

For purpose of reduction of these errors, a feedback from the visual motion compensation is used. Represented by the difference between the reference and stabilized images, it indicates regions where the discrete physical model is inaccurate. For increasing the spatial resolution of this model, additional model nodes are inserted in these regions. For illustrative purposes, these nodes are denoted in Fig. 2.3 by green points. Furthermore, the material properties of the model are refined in these regions by assigned additional model parameters to the inserted nodes.

For purpose of adapting the model parameters, a simultaneous state and parameter estimation involved in a tracking method is employed. It provides the estimates of the model parameters along with the system state by processing camera measurements extracted from incoming camera images. In addition to, the model parameters assigned to the inserted model nodes are corrected in this way. As a result, the discretization as well as material properties of the discrete physical model are continuously adapted only where necessary according to the feedback provided by the visual motion compensation. In this way, the material properties of the model become inhomogeneous.

As refinement of the discretization not only improves the accuracy but also increases the degrees of freedom of the discrete model, this adaptation strategy allows us to achieve high accuracy of the model by a tractable computational complexity. The unique part of the adaptation, is that not only the discrete physical model is permanently adjusted to the object's behavior, but also all physics-based models that are derived from it.

On the whole, a continuous monitoring of the quality of the system based on feedback from a visual motion compensation and runtime improvement of the models has the advantages of having lower times for system initialization. Furthermore, the models are flexibly adapted to different objects and changing conditions.

2.4 Summary

In this chapter, three main ideas for tracking and visual motion compensation are proposed. The first idea is based upon the physical interpretation of both methods. The second idea is motivated by the aim to get insight into the complex behavior of the object under observation. For this purpose, the object's behavior is approximated by a physical model, which approximates the object's motion directly in a physical space. In this way, a physically correct estimation of the object's position as well as a physically correct reconstruction of the camera images showing this object are ensured. The third idea consists of a runtime adaptation of all mathematical models based upon a feedback from the visual motion compensation and simultaneous state and parameter estimation. As a result, the quality of the entire system is continuously evaluated and improved.

3 Probabilistic Models for Distributed Parameter Systems

With the aim of providing appropriate tools for heart motion reconstruction, a discrete state-space model is established in this section. For this purpose, at first, the physical model of the heart wall is deduced based upon physical principles capturing the motion of the heart. This model describes the space-time continuous behavior of the heart by a distributed parameter system, which state depends not only on time but also on spatial coordinates. The mathematical formulation of this system yields a system of stochastic partial differential equations[2]. As schematically illustrated in Fig. 3.1, this system of equations serves as a starting point for derivation of the state-space model consisting of the system and measurement models.

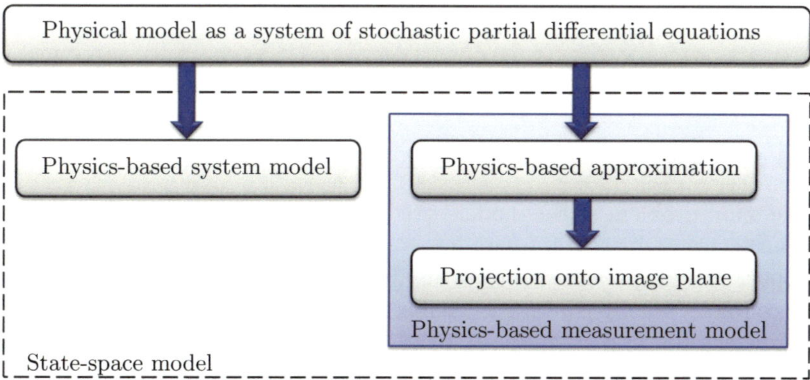

Figure 3.1: In this chapter, the physics-based system and measurement models are derived from the physical model that is mathematically formulated in the form of a system of stochastic partial differential equations.

---

[2]In this thesis, the distributed parameter system is described by the system of stochastic partial differential equations with uncertain input, random initial and boundary conditions. The randomness of the coefficients of this equation is considered after the conversion of the system into a discrete form.

The physics-based system model describes the temporal propagation of the system state. In order to obtain this model, the system of stochastic partial differential equations is discretized in space using efficient element-free method [12–14], yielding the system of stochastic ordinary differential equations, which is solved by a suitable time integration method.

The physics-based measurement model establishes the relationship between the system state and measurements extracted from incoming camera images. It is obtained by projecting the numerical solution of the system of stochastic partial differential equations, which approximates current position of the heart wall, in the camera view. This solution is provided by a physics-based approximation, the parameters of which encompass the physical information incorporated in a system of stochastic partial differential equations.

Partly, the physical model of the heart wall as well as the derivation of the state-space model have been presented in our papers [238, 240, 241]. This chapter extends these publications in particular by the theoretical aspects concerning the derivation of the physical model from physical principles underlying the motion of the object. Furthermore, the extension of the model with respect to Kelvin-Voigt viscoelastic material represents unpublished material.

3.1 Problem Formulation

The main problem of the model-based processing of measurement data is to achieve a high accuracy with low computational complexity. Thus, with respect to the application in beating heart surgery, the heart model must be computationally efficient to enable runtime tracking and visual motion compensation. Furthermore, it must be sophisticated to reconstruct the motion of the heart accurately enough.

Among all categories of the methods presented in Section 1.3, the physics-based methods are superior in accuracy [42]. Based upon an appropriate physical model, such methods guarantee physical correctness of the reconstruction due to an incorporation of a large amount of a priori knowledge. This is especially advantageous when no measurement information is available or when this information is sparse or even corrupted by outliers. Therefore, for accurate reconstruction of the heart motion, the complex behavior of the heart must be approximated by a physical model. Unfortunately, such models are quite complex. That hinders their application in a beating heart surgery system, where the capability of real-time operability is essential.

As a matter of fact, the physical model, even when it is highly realistic, cannot exactly represent the object under observation due to unknown object's behavior, noisy excitation or numerical errors. Therefore, the description of accompanied systematic and stochastic uncertainties should be incorporated in the model. An appropriate mathematical description of such a probabilistic model is a system of stochastic partial differential equations that defines the object displacement as a random field consisting of a set of spatially and temporally distributed random variables.

As the physical model and the camera measurements are corrupted by uncertainties, the position of the object should be estimated using stochastic filters. For this purpose, the system of stochastic partial differential equations describing the physical model must be converted in a state-space form. Because of the intractability of this problem [116], the numerical methods should be applied for converting the system in a discrete state-space model. This model consists of a discrete system model and a discrete measurement model given by equations

$$
\begin{aligned}
\underline{z}_{k+1} &= \underline{a}_k\left(\underline{z}_k, \hat{\underline{u}}_k, \underline{s}_k, \underline{w}_k^z\right), \\
\hat{\underline{y}}_k &= \underline{h}_k\left(\underline{z}_k, \underline{e}_k, \underline{v}_k\right).
\end{aligned}
\tag{3.1}
$$

Here, system function $\underline{a}_k : \mathbb{R}^{n_z} \times \mathbb{R}^{n_u} \times \mathbb{R}^{n_s} \times \mathbb{R}^{n_{wz}} \to \mathbb{R}^{n_z}$, $k \in \mathbb{N}$ propagates the system state $\underline{z}_k \in \mathbb{R}^{n_z}$ from time step t_k to time step t_{k+1} by processing the known excitation $\hat{\underline{u}}_k \in \mathbb{R}^{n_{\hat{u}}}$. The systematic and stochastic errors of the system model are denoted by $\underline{s}_k \in \mathbb{R}^{n_s}$ and $\underline{w}_k^z \in \mathbb{R}^{n_{wz}}$ respectively. The random vector $\underline{w}_k^z$ characterizes the system noise that is assumed to be zero-mean white. The measurement function $\underline{h}_k : \mathbb{R}^{n_z} \times \mathbb{R}^{n_e} \times \mathbb{R}^{n_v} \to \mathbb{R}^{n_y}$ relates measurements $\hat{\underline{y}}_k \in \mathbb{R}^{n_y}$ to the system state $\underline{z}_k$ at the current time step. The systematic and stochastic errors of the measurement model are denoted by $\underline{e}_k \in \mathbb{R}^{n_e}$ and $\underline{v}_k \in \mathbb{R}^{n_v}$. The random vector $\underline{v}_k$ describing the measurement noise is assumed to be zero-mean white and independent of the system noise term $\underline{w}_k^z$.

Frequently, for conversion of a system of stochastic partial differential equations into a discrete state-space form, numerical methods for spatial and temporal discretization are employed. The main point here is to choose a proper numerical method that is most suitable for the solution of this problem, i.e., a method that is computationally efficient and sufficiently accurate.

3.2 Physical Model of Deformable Object

A novel physical model of the heart wall, suitable for beating heart surgery system, is presented in this section. It is formulated in form of a system of stochastic partial differential equations. The first step consists in a description of the heart motion by a system of deterministic equations, which is derived from an energy conservation principle. This principle sets the requirement on the stationarity of the total energy of a closed dynamic system [53], where no energy can be added or lost. Thereafter, these equations are extended by damping forces, which account for the occurred energy dissipation. Finally, the uncertainties arising due to unknown input forces and initial conditions are incorporated in the motion equations. For further details to the elasticity theory used for derivation of the model, we refer to the textbooks [11, 53, 91].

Physical Description of Deformable Objects

This section deals with the approximation of the heart wall by a physical deformable body subjected to external forces. Before describing in detail the geometry and material characteristics of this body, firstly, the functionality of the heart is briefly presented, following which the simplifications and assumptions that are made on the heart behavior are introduced.

Objects under Consideration   In this paragraph, we will briefly look into the structure and functionality of the heart with the aim to discuss the properties of the heart that are essential for the modeling of the heart wall. A detailed description of the heart can be found in textbooks [93, 108, 201, 225].

The human heart consisting of four chambers – two atria and two ventricles – is responsible for pumping blood to the entire body. This is a hollow muscular organ enclosed in the membranous sac called a pericardium. This sac can be exposed by a surgeon during beating heart operations for gaining direct access to the heart.

The heart itself is supplied by oxygenated blood delivered by coronary arteries. Any narrowing, or blockage, of these arteries, e.g., due to atherosclerosis, results in a coronary artery disease [201] that impairs the blood supply to the heart, causes an abnormality of cardiac function, and in end-stage, the heart failure. The most common surgical treatment of this disease is the restoration of the blood supply by the coronary artery bypass grafting, which is often performed on a beating heart.

We will focus on beating heart surgical interventions carried out on the **left ventricle**, because this chamber is mostly affected by the coronary artery disease. The reason for this relates to the harder work of the left ventricle than that of all other chambers. Its work consists in ejecting blood from the heart into the major circulatory network.

The size as well as physical properties of the left ventricle differ from person to person and strongly depend on whether the person suffers from atherosclerosis or not. The wall of the ventricle is of a **special kind of muscle**, called myocardium, that is composed of several spirally wrapped layers. According to medical statistical studies [62, 187], the thickness of the ventricular wall can vary from 6 mm up to 20 mm depending on the person.

Electrical activity, known as a cardiac cycle, regulates the rhythmic motion of the heart consisting of two phases, called systole and diastole. During diastole, two atria excited by an electrical impulse are forced to contract. This leads to the lower pressure in the ventricle than that in the atria. As a result, the antrioventricular valves open and blood from the atria refills the relaxed ventricles. During systole, the myocardium contracts due to stimulation by electrical currents. The tension of the myocardium leads to the increase of the pressure inside the ventricle, so that tricuspid and mitral valves, separating the ventricles from atria, close. Then, when pulmonary and aortic valves open due to sufficient pressure inside of the ventricles, blood is forced into the pulmonary artery and the aorta.

It should be noted that **electrical currents** generated in cells of the myocardium [201] are distributed over the heart muscle through muscle fibers.

Simplifications and Assumptions For purpose of achieving a high efficiency of the model, only the **part of the heart wall** of the left ventricle is considered that is important for beating heart operations on coronary arteries. This is the area on which the surgeon is working, as schematically illustrated in Fig. 3.2.

Normally, the area under consideration is **stabilized** during beating heart operations. For this purpose, for example, a vacuum-based mechanical

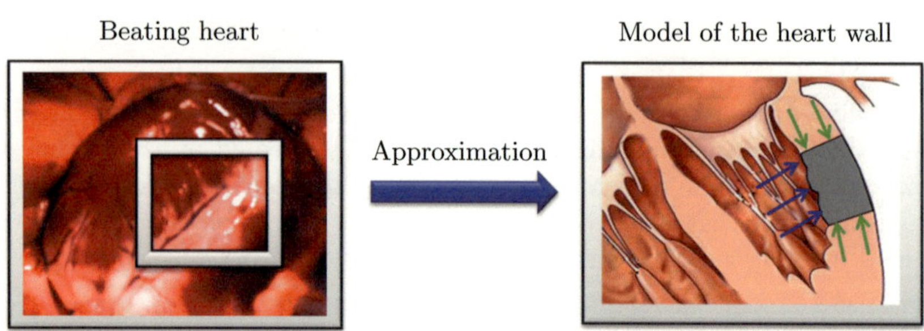

Figure 3.2: Approximation of the heart wall of the left ventricle by a heart model. The model considers the heart wall inside the operation area. This area is subjected to the external forces, such as pressure forces inside of the cardiac chamber (blue) and muscle forces (green).

stabilizer, such as Octopus [43,61,122] can be employed. As stated in [43], the amplitude of the remaining cardiac motion after stabilization, can achieve at some points about 0.3 ± 0.33 mm in x and y directions and about 2 ± 2.6 mm in the z direction.

For modeling the complex behavior of this area, some assumptions are made on the physical properties of the heart wall and cardiac cycle. First of all, the complex layered structure of the myocardium is simplified. Because of the small size of the considered area, primarily, it is supposed that the heart tissues inside of this area consist of **isotropic and homogeneous material**. This means that the properties of the material are independent of direction and position. However, since the myocardium is heterogeneous by nature, the material properties of the model are further adapted after the discretization of the model with regard to **material inhomogeneity**. It should be noted that the subsequent adaptation of the material properties allows to avoid the substantial complication of the model. The material of the heart wall is assumed to be sufficiently smooth, with **damping properties**, caused, e.g., by an energy conversion to the heat or inner damping. Furthermore, the heart wall is supposed to have **constant thickness** within the stabilized area.

The behavior of the heart tissues is assumed to be **linear viscoelastic**. This means that the stress-strain response is reversible and the heart wall returns in an undeformed configuration in the absence of loads. In this configuration, its material is unstressed. Conversely, applied loads cause non-zero stresses, as well as strains in the material and hence, force the object to occupy a deformed configuration. This linear assumption is justified by the fact that the displacements of the mechanically stabilized heart are small.

The viscoelastic description of the heart muscle allows for the reproduction of such tissue properties where an elastic model would fail. For example, when the tissue is deformed and its deformation is maintained, the inner forces of material or stress will diminish with time. This property is called stress relaxation [9]. Furthermore, when the tissue is subjected to constant external force, its deformation will still continue to increase with time. This phenomenon is characterized by strain retardation [9]. Moreover, the strain-stress response of the object, subjected to the repeated load and relaxation, will be deemed to be as a hysteresis, due to dissipation of mechanical energy.

With regard to the cardiac cycle, it should be noted that, although the heart motion is initiated by forces generated in a distributed fashion within the heart muscles that are stimulated by electrical currents, it is assumed that the motion of the stabilized area is caused by a **pressure** inside the myocardium. This assumption is justified by the medical experimental studies [191, 192], which state that the active force generated by the myocardium can be reproduced by ventricular pressure. The pressure inside the left ventricle can be measured by a heart catheter or echocardiography [158].

Model Geometry　For the definition of the model geometry, a fixed time-independent Cartesian coordinate system in $\mathbb{R}^3$ with origin in O and x, y, z axes is introduced, as depicted in Fig. 3.3, where the model of the heart wall within the stabilized area is schematically illustrated. The mechanical equilibrium of the heart wall (also called undeformed configuration) is described by the bounded domain $\overline{\Omega} \subseteq \mathbb{R}^3$. This domain $\overline{\Omega} = \Omega \cup \partial\Omega$ contains the interior Ω and smooth boundary $\partial\Omega$. The boundary $\partial\Omega = \Gamma_N \cup \Gamma_R$ is decomposed in two parts: Neumann boundary Γ_N, where pressure forces act, and Robin boundary Γ_R that models the connection of the considered heart wall with the surrounding heart tissues. One can imagine this connection as damped springs of stiffness β, as shown in Fig. 3.3. The wall of the heart is assumed to be affected not only by pressure forces $\underline{f}_N : \Gamma_N \times \mathbb{R}_+ \to \mathbb{R}^3$ but also by muscle strengths $\underline{f}_R : \Gamma_R \times \mathcal{I} \to \mathbb{R}^3$, and body forces per unit volume $\underline{f}_I : \Omega \times \mathcal{I} \to \mathbb{R}^3$, which are defined by

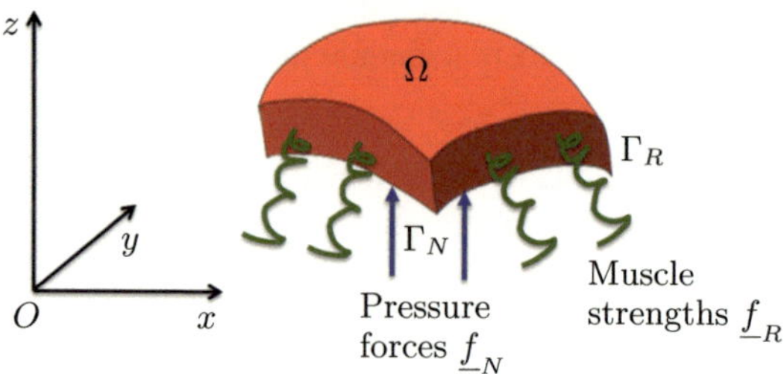

Figure 3.3: Geometry of the heart model, whereby the connection with the surrounding heart tissues is described by the Robin boundary Γ_R. Pressure forces $\underline{f}_N$ act on Neumann boundary Γ_N.

continuous vector functions depending on the spatial coordinates $\underline{r} \in \overline{\Omega}$ and time $t \in \mathcal{I} \subseteq \mathbb{R}_+$ in the interval $\mathcal{I} = [0, t_k]$ up to time step t_k. The body forces are not indicated in Fig. 3.3, because they act on the entire object. One example of such force is the gravitational force.

Physics-Based Motion Description

For the derivation of the mathematical model of the heart wall behavior based upon the energy conservation principle, in this section, the kinetic and the potential energies of the heart wall are determined by presuming the linear elastic behavior of the heart. The elastic assumption on the heart muscle will be later replaced with viscoelastic.

At first, the motion of the heart wall is linearized around the undeformed configuration of the object defined by the domain $\overline{\Omega} \subseteq \mathbb{R}^3$, as shown in Fig. 3.4. The space-time continuous deformation of the object is then mathematically described by a continuous mapping $\underline{\psi} : \overline{\Omega} \times \mathcal{I} \to \overline{\Omega}^\psi$ of the domain $\overline{\Omega}$ to its deformed configuration $\overline{\Omega}^\psi = \underline{\psi}\left(\overline{\Omega}, t\right)$ with interior Ω^ψ and smooth boundary $\partial\Omega^\psi$ at time $t \in \mathcal{I} \subseteq \mathbb{R}_+$ from the interval $\mathcal{I}$. The points $\underline{r} = [x^r, y^r, z^r]^\mathrm{T} \in \overline{\Omega}$ in the domain $\overline{\Omega}$ are assumed to be uniquely assigned to the points $\underline{r}^\psi = [x^{r^\psi}, y^{r^\psi}, z^{r^\psi}]^\mathrm{T} \in \overline{\Omega}^\psi$ in the domain $\overline{\Omega}^\psi \subseteq \mathbb{R}^3$ at every time $t \in \mathcal{I}$ by

$$\underline{r}^\psi = \underline{\psi}\left(\underline{r}, t\right) = \underline{r} + \underline{d}\left(\underline{r}, t\right) , \tag{3.2}$$

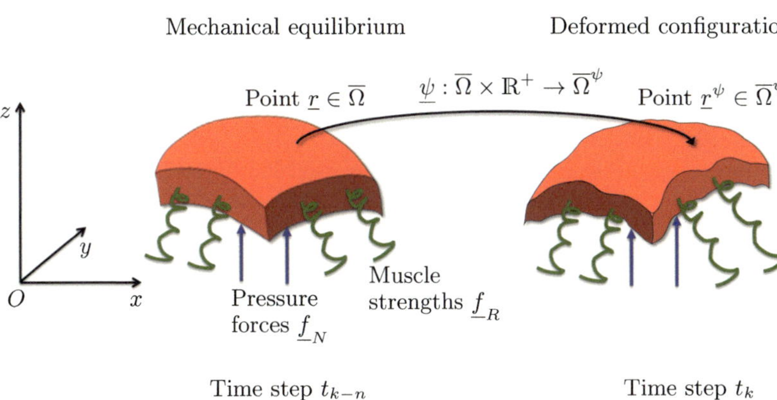

Figure 3.4: Mathematical description of the heart motion by the continuous function $\underline{\psi}$, which maps the undeformed configuration of the model $\overline{\Omega}$ to its deformed configuration $\overline{\Omega}^\psi = \underline{\psi}\left(\overline{\Omega}, t\right)$. The model is affected by muscle strengths $\underline{f}_R$ and pressure forces $\underline{f}_N$.

where the continuous vector function $\underline{d} : \overline{\Omega} \times \mathcal{I} \to \mathcal{D}$ is assumed to be twice differentiable with respect to time and sufficiently smooth in space and time. Its values denote the displacement field in space $\mathcal{D} \subseteq \mathbb{R}^3$.

Strain as a Measure of Deformation For description of the object's deformation, the strain of the heart tissues $\mathcal{E}$, which describes the changes in length elements compared to their original length, is defined by a second order tensor in matrix representation

$$\mathcal{E} = \{\epsilon_{ij}\}, \ 1 \leq i \leq 3, \ 1 \leq j \leq 3. \tag{3.3}$$

In geometrical interpretation, the diagonal elements of the strain tensor are relative elongations of the line elements along coordinate axes, whereby the off-diagonal elements are related to the shear angles [11].

As the assumption on the linear elastic behavior of the object under consideration implies small deformations of the heart wall, the strain can be approximated by Cauchy's strain tensor

$$\mathcal{E}\left(\underline{d}\left(\underline{r}, t\right)\right) = \frac{1}{2}\left(\frac{\partial \underline{d}\left(\underline{r}, t\right)}{\partial \underline{r}} + \left(\frac{\partial \underline{d}\left(\underline{r}, t\right)}{\partial \underline{r}}\right)^{\mathrm{T}}\right), \tag{3.4}$$

where the deformation gradient defined by the derivative of the function providing the displacement field $\underline{d} = [x^d, y^d, z^d]^{\mathrm{T}} \in \mathcal{D}$ is described by the Jacobian matrix

$$\frac{\partial \underline{d}\left(\underline{r}, t\right)}{\partial \underline{r}} = \begin{bmatrix} \frac{\partial x^d}{\partial x^r} & \frac{\partial x^d}{\partial y^r} & \frac{\partial x^d}{\partial z^r} \\ \frac{\partial y^d}{\partial x^r} & \frac{\partial y^d}{\partial y^r} & \frac{\partial y^d}{\partial z^r} \\ \frac{\partial z^d}{\partial x^r} & \frac{\partial z^d}{\partial y^r} & \frac{\partial z^d}{\partial z^r} \end{bmatrix} .$$

This tensor is the linearization of the nonlinear Green and St. Venant strain tensor [53], so that both of the tensors coincide in the presence of small deformations. The evaluation of Cauchy's strain tensor is cheaper, because it is described by a linear function of displacements. However, the drawback of this tensor is that it is not rotationally invariant. In order to avoid poor results in presence of rotations, a corotational formulation of the Cauchy's strain tensor [25, 148] can be used. The key idea of this formulation is to determine the rotation of each point of the object by exploiting the decomposition of the deformation gradient into a rotational part and a stretching part. In this thesis, it is assumed that the heart wall within the stabilized area is not subjected to rotations because of its immobilization by means of a mechanical stabilizer. This permits the use of the Cauchy's strain tensor in its original form. However, it should be noted that for the beating heart without mechanical stabilization, this assumption may be inadequate, because the non-stabilized heart undergoes large rotations [189].

Stress as a Description of Inner Forces   The inner forces of the deformable object hold the object in its shape [11] and act against the applied loads. These forces are described by the stress that is defined as an inner force taken in a unit area. The stress of the heart tissues is here approximated according to Hooke's law [53] in the form

$$\boldsymbol{\Sigma}\left(\underline{d}\left(\underline{r}, t\right)\right) = \lambda \operatorname{trace}\left(\boldsymbol{\mathcal{E}}\left(\underline{d}\left(\underline{r}, t\right)\right)\right)\mathbf{I} + 2\mu \boldsymbol{\mathcal{E}}\left(\underline{d}\left(\underline{r}, t\right)\right), \tag{3.5}$$

where $\mathbf{I} \in \mathbb{R}^{3 \times 3}$ is the identity matrix. As follows from this equation, the stress tensor $\boldsymbol{\Sigma}$ denoted by the matrix

$$\boldsymbol{\Sigma} = \{\sigma_{ij}\}, \ 1 \leq i \leq 3, \ 1 \leq j \leq 3 \tag{3.6}$$

reflects the physical properties of the object defined by Lamé constants

$$\lambda = \frac{E\nu}{\left(1 + \nu\right)\left(1 - 2\nu\right)}, \ \mu = \frac{E}{2\left(1 + \nu\right)}, \tag{3.7}$$

material density ρ, modulus of elasticity also called Young modulus E, and Poisson ratio ν, which is defined as ratio of lateral strain to axial strain.

Naturally, Hooke's law describes the stress as a linear function of the strain. This is only valid for small displacements of the object. Furthermore, due to assumption on the isotropy of the heart wall material, the material properties, specially Poisson ratio ν, Young modulus E, and second Lamé constant μ known as shear modulus, are not independent. If two of them are known, the third one can be obtained according to (3.7). Although the anisotropic description of the heart tissues reproduces better the layered structure of the myocardium, it would significantly increase the number of the material parameters. While an isotropic material is described by only two constants ν and E, we need 21 constants for an anisotropic material. For a material with orthogonal directions of anisotropy, so called orthotropic material, the number of these constants reduces to 6. Therefore, an isotropic material is chosen with the aim of minimal parametrization of the heart model.

Energy of the Heart Wall   In the following, the total energy of the heart wall is deduced according to the principle of energy conservation, which states that the total energy of a closed mechanical system remains constant over time. It should be noted that the mechanical system is closed when no energy may be added or lost. In this case, the energy can be only transformed from one form to another.

Here, we assume that there is no changes in the total energy of the heart wall, in order to allow for the variational formulation, which is based on the principle of energy conservation. Later, this assumption will be abolished by the generalization of the Hooke's law for energy dissipation.

Accordingly, since the energy of the heart wall is supposed to change from kinetic to potential, the total energy is determined by the difference between these two types of energies

$$\mathbb{W}\left[\underline{d}\left(\underline{r},t\right)\right] = \mathbb{K}\left[\underline{d}\left(\underline{r},t\right)\right] - \mathbb{U}\left[\underline{d}\left(\underline{r},t\right)\right], \tag{3.8}$$

where the kinetic energy of the heart wall is described by the functional

$$\mathbb{K}\left[\underline{d}\left(\underline{r},t\right)\right] = \frac{1}{2}\int_{\Omega} \rho\left(\underline{\dot{d}}\left(\underline{r},t\right)\right)^{\mathrm{T}} \underline{\dot{d}}\left(\underline{r},t\right)\mathrm{d}\underline{r}.$$

A superimposed dot denotes here differentiation with respect to time t.

For deducing the potential energy, the Frobenius inner product of the stress and strain tensors is further defined by the trace of their inner product

$$\boldsymbol{\Sigma}\left(\underline{d}\left(\underline{r},t\right)\right) : \boldsymbol{\mathcal{E}}\left(\underline{d}\left(\underline{r},t\right)\right) := \text{trace}\left(\left(\boldsymbol{\Sigma}\left(\underline{d}\left(\underline{r},t\right)\right)\right)^{T}\boldsymbol{\mathcal{E}}\left(\underline{d}\left(\underline{r},t\right)\right)\right),$$

where the symbol := has the meaning "is defined by". Then, the potential energy involving the internal energy, which is influenced by the work of the stiffness and surface forces, as well as by the work of muscle strengths and body forces is determined by

$$\mathbb{U}\left[\underline{d}\left(\underline{r},t\right)\right] = \underbrace{\frac{1}{2}\int_{\Omega}\boldsymbol{\Sigma}\left(\underline{d}\left(\underline{r},t\right)\right) : \boldsymbol{\mathcal{E}}\left(\underline{d}\left(\underline{r},t\right)\right)\mathrm{d}\underline{r}}_{\text{internal energy}} + \underbrace{\frac{1}{2}\int_{\Gamma_R}\beta\left(\underline{d}\left(\underline{r},t\right)\right)^{T}\underline{d}\left(\underline{r},t\right)\mathrm{d}\underline{r}}_{\text{work of stiffness forces}}$$

$$-\underbrace{\int_{\Omega}\left(\underline{d}\left(\underline{r},t\right)\right)^{\mathrm{T}}\underline{f}_{I}\left(\underline{r},t\right)\mathrm{d}\underline{r}}_{\text{work of body forces}} - \underbrace{\int_{\Gamma_N}\left(\underline{d}\left(\underline{r},t\right)\right)^{\mathrm{T}}\underline{f}_{N}\left(\underline{r},t\right)\mathrm{d}\underline{r}}_{\text{work of pressure forces}}$$

$$-\underbrace{\int_{\Gamma_R}\left(\underline{d}\left(\underline{r},t\right)\right)^{\mathrm{T}}\underline{f}_{R}\left(\underline{r},t\right)\mathrm{d}\underline{r}}_{\text{work of muscle strengths}},$$

where the parameter β stands for the stiffness of the heart tissues on the Robin boundary.

As a result, the total energy obtained by plugging the kinetic and potential energy in (3.8), reveals the fact that the work done by external forces is equal to the change of the internal energy.

Variational Formulation

The total energy of the heart wall is used in this section for the derivation of the mathematical model of the heart wall behavior. This model will be expressed by the variational formulation obtained by applying the calculus of variations [200].

According to the principle of stationary action [115], the displacement field $\underline{d}$, introduced in (3.2), is characterized by the requirement, which states that the total energy

$$\int_{t_{k-1}}^{t_k} \mathrm{W}\left[\underline{d}\left(\underline{r}, t\right)\right] \mathrm{d}t \tag{3.9}$$

stored between two arbitrary time steps t_{k-1} and t_k must have a stationary value, not necessarily a minimum. In accordance with the calculus of variations [200], this means that the first variation of the integral (3.9) must vanish.

In order to compute the first variation, it is supposed that the displacement $\underline{d} \in \mathcal{D}$ of the point $\underline{r} \in \overline{\Omega}$ in the domain $\overline{\Omega}$ is perturbed between the time steps t_{k-1} and t_k. These perturbations can be described by all admissible functions $\underline{v} : \overline{\Omega} \times \mathbb{R}_+ \to \mathbb{R}^{n_v}$. The functions are admissible when their values $\underline{v}\left(\cdot, t_{k-1}\right) = \underline{v}\left(\cdot, t_k\right) = 0$ are zero at the time steps t_{k-1} and t_k and small at other time steps. Then, the displacement $\underline{d} \in \mathcal{D}$ is the stationary solution of the functional

$$\mathrm{V}\left[\underline{d}\left(\underline{r}, t\right) + \theta \underline{v}\left(\underline{r}, t\right)\right] = \int_{t_{k-1}}^{t_k} \mathrm{W}\left[\underline{d}\left(\underline{r}, t\right) + \theta \underline{v}\left(\underline{r}, t\right)\right] \mathrm{d}t, \tag{3.10}$$

where $\theta \in \mathbb{R}$ is the real number. Consequently, the stationary solution is defined by a directional, also called Gâteaux [200] derivative, of the functional V at $\underline{d}$ in the direction $\underline{v}$ according to

$$\left. \frac{\mathrm{d}}{\mathrm{d}\theta} \mathrm{V}\left[\underline{d}\left(\underline{r}, t\right) + \theta \underline{v}\left(\underline{r}, t\right)\right] \right|_{\theta=0} = 0. \tag{3.11}$$

The substitution of (3.8) in (3.10) with the subsequent calculation of the Gâteaux derivative according to (3.11) leads to

$$\int_{t_{k-1}}^{t_k} \left(\int_{\Omega} \rho\left(\underline{\dot{v}}\left(\underline{r}, t\right)\right)^{\mathrm{T}} \underline{\ddot{d}}\left(\underline{r}, t\right) \mathrm{d}\underline{r} - \int_{\Omega} \boldsymbol{\Sigma}\left(\underline{d}\left(\underline{r}, t\right)\right) : \boldsymbol{\mathcal{E}}\left(\underline{v}\left(\underline{r}, t\right)\right) \mathrm{d}\underline{r} \right.$$
$$- \int_{\Gamma_R} \beta\left(\underline{v}\left(\underline{r}, t\right)\right)^{\mathrm{T}} \underline{\dot{d}}\left(\underline{r}, t\right) \mathrm{d}\underline{r} + \int_{\Omega} \left(\underline{v}\left(\underline{r}, t\right)\right)^{\mathrm{T}} \underline{f}_I\left(\underline{r}, t\right) \mathrm{d}\underline{r} \tag{3.12}$$
$$\left. + \int_{\Gamma_N} \left(\underline{v}\left(\underline{r}, t\right)\right)^{\mathrm{T}} \underline{f}_N\left(\underline{r}, t\right) \mathrm{d}\underline{r} + \int_{\Gamma_R} \left(\underline{v}\left(\underline{r}, t\right)\right)^{\mathrm{T}} \underline{f}_R\left(\underline{r}, t\right) \mathrm{d}\underline{r} \right) \mathrm{d}t = 0.$$

In the following, (3.12) is decomposed in order to obtain the variational equation. Hence, the partial integration of the first integral with the respect to time results in

$$
\int_{t_{k-1}}^{t_k} \int_{\Omega} \rho \left(\dot{\underline{v}} \left(\underline{r}, t \right) \right)^{\mathrm{T}} \dot{\underline{d}} \left(\underline{r}, t \right) \mathrm{d}\underline{r} \, \mathrm{d}t = \int_{\Omega} \rho \left(\underline{v} \left(\underline{r}, t \right) \right)^{\mathrm{T}} \dot{\underline{d}} \left(\underline{r}, t \right) \mathrm{d}\underline{r} \Bigg|_{t_{k-1}}^{t_k}
$$

$$
- \int_{t_{k-1}}^{t_k} \int_{\Omega} \left(\underline{v} \left(\underline{r}, t \right) \right)^{\mathrm{T}} \rho \ddot{\underline{d}} \left(\underline{r}, t \right) \mathrm{d}\underline{r} \, \mathrm{d}t \,,
$$

where the first term vanishes because the function values $\underline{v} \left(\cdot, t_{k-1} \right) = \underline{v} \left(\cdot, t_k \right)$ coincide at both time steps. Furthermore, the second integral, which defines the variation of the strain energy, is decomposed by exploiting Green's formula [33] according to

$$
\int_{\Omega} \boldsymbol{\Sigma} \left(\underline{d} \left(\underline{r}, t \right) \right) : \boldsymbol{\mathcal{E}} \left(\underline{v} \left(\underline{r}, t \right) \right) \mathrm{d}\underline{r} = \int_{\Gamma_N} \left(\underline{v} \left(\underline{r}, t \right) \right)^{\mathrm{T}} \boldsymbol{\Sigma} \left(\underline{d} \left(\underline{r}, t \right) \right) \underline{n} \, \mathrm{d}\underline{r}
$$

$$
+ \int_{\Gamma_R} \left(\underline{v} \left(\underline{r}, t \right) \right)^{\mathrm{T}} \boldsymbol{\Sigma} \left(\underline{d} \left(\underline{r}, t \right) \right) \underline{n} \, \mathrm{d}\underline{r} - \int_{\Omega} \left(\underline{v} \left(\underline{r}, t \right) \right)^{\mathrm{T}} \mathrm{div} \, \boldsymbol{\Sigma} \left(\underline{d} \left(\underline{r}, t \right) \right) \mathrm{d}\underline{r} \,,
$$

where the divergence of the tensor $\boldsymbol{\Sigma} \in \mathbb{R}^{3 \times 3}$ is defined as the vector with components that are the divergences of the rows of the tensor

$$
\mathrm{div} \, \boldsymbol{\Sigma} \left(\underline{d} \left(\underline{r}, t \right) \right) = \begin{bmatrix} \frac{\partial \sigma_{11}}{\partial x^r} + \frac{\partial \sigma_{12}}{\partial y^r} + \frac{\partial \sigma_{13}}{\partial z^r} \\ \frac{\partial \sigma_{21}}{\partial x^r} + \frac{\partial \sigma_{22}}{\partial y^r} + \frac{\partial \sigma_{23}}{\partial z^r} \\ \frac{\partial \sigma_{31}}{\partial x^r} + \frac{\partial \sigma_{32}}{\partial y^r} + \frac{\partial \sigma_{33}}{\partial z^r} \end{bmatrix} \,. \tag{3.13}
$$

As a result, (3.12) is converted into the variational equation

$$
\int_{t_{k-1}}^{t_k} \Bigg(\int_{\Omega} \left(\underline{v} \left(\underline{r}, t \right) \right)^{\mathrm{T}} \left(\rho \ddot{\underline{d}} \left(\underline{r}, t \right) - \mathrm{div} \, \boldsymbol{\Sigma} \left(\underline{d} \left(\underline{r}, t \right) \right) - \underline{f}_I \left(\underline{r}, t \right) \right) \mathrm{d}\underline{r}
$$

$$
+ \int_{\Gamma_N} \left(\underline{v} \left(\underline{r}, t \right) \right)^{\mathrm{T}} \left(\boldsymbol{\Sigma} \left(\underline{d} \left(\underline{r}, t \right) \right) \underline{n} - \underline{f}_N \left(\underline{r}, t \right) \right) \mathrm{d}\underline{r} \tag{3.14}
$$

$$
+ \int_{\Gamma_R} \left(\underline{v} \left(\underline{r}, t \right) \right)^{\mathrm{T}} \left(\boldsymbol{\Sigma} \left(\underline{d} \left(\underline{r}, t \right) \right) \underline{n} + \beta \underline{d} \left(\underline{r}, t \right) - \underline{f}_R \left(\underline{r}, t \right) \right) \mathrm{d}\underline{r} \Bigg) \mathrm{d}t = 0 \,.
$$

It should be noted that this equation must hold for all admissible functions $\underline{v}$ (set, e.g., $\underline{v}$=const.). Hence, the integrals in the parenthesis must be equal to zero. This is fulfilled when the object is in equilibrium, i.e., the inner forces of the object are canceled by the imposed external forces, like, e.g., body forces $\underline{f}_I$, pressure forces $\underline{f}_N$, and muscle strengths $\underline{f}_R$, at every point $\underline{r} \in \overline{\Omega}$ and time step $t \in \mathcal{I} \subseteq \mathbb{R}^3$.

Energy Dissipation

Since the variational formulation is deduced based on the energy conservation principle, it cannot cover the energy dissipation, e.g., due to conversion of energy to the heat or inner damping of the material. However, due to energy dissipation, the heart wall, subjected to an impulse force, will occupy its undeformed configuration after a while, instead of performing periodic motion. Hence, in this section, the variational equation (3.14) is extended for the damping forces. This results in a viscoelastic representation of the heart tissues material, for which the stress-strain response depends on the strain rate.

The damping properties of the material are modeled by a Rayleigh damping introduced in [171]. By assuming that the energy dissipation is proportional to the material stiffness and density, this model has the advantage that for just two modes the damping ratios need to be specified. These ratios are defined by the coefficient γ describing the damping proportional to stiffness and the coefficient κ denoting the damping proportional to the density.

To take into account the Rayleigh damping, the variational equation (3.14) is modified according to the theoretical framework proposed in [91]. In this context, the damping effect proportional to the stiffness is introduced by a generalized Hooke's law

$$
\begin{aligned}
\boldsymbol{\Sigma}\left(\underline{d}(\underline{r},t),\underline{\dot{d}}(\underline{r},t)\right) = {}& \lambda \operatorname{trace}\left(\boldsymbol{\mathcal{E}}\left(\underline{d}(\underline{r},t) + \gamma\underline{\dot{d}}(\underline{r},t)\right)\right)\mathbf{I} \\
& + 2\mu\left(\boldsymbol{\mathcal{E}}\left(\underline{d}(\underline{r},t) + \gamma\underline{\dot{d}}(\underline{r},t)\right)\right),
\end{aligned}
\tag{3.15}
$$

which now defines the stress tensor from (3.5). In this way, it is presumed that the heart tissue is sufficiently accurate represented by Kelvin-Voigt viscoelastic material [9], which is modeled by a parallel connection of a spring and a dashpot.

In the next step, the damping force proportional to the material density is introduced in the variational equation (3.14). This results in

$$
\int_{\Omega} (\underline{v}\,(\underline{r},t))^{\mathrm{T}} \left(\rho \underline{\ddot{d}}\,(\underline{r},t) + \rho \kappa \underline{\dot{d}}\,(\underline{r},t) - \operatorname{div} \boldsymbol{\Sigma}\left(\underline{d}\,(\underline{r},t), \underline{\dot{d}}\,(\underline{r},t) \right) - \underline{f}_I\,(\underline{r},t) \right) \mathrm{d}\underline{r}
$$

$$
+ \int_{\Gamma_N} (\underline{v}\,(\underline{r},t))^{\mathrm{T}} \left(\boldsymbol{\Sigma}\left(\underline{d}\,(\underline{r},t), \underline{\dot{d}}\,(\underline{r},t) \right) \underline{n} - \underline{f}_N\,(\underline{r},t) \right) \mathrm{d}\underline{r}
$$

$$
+ \int_{\Gamma_R} (\underline{v}\,(\underline{r},t))^{\mathrm{T}} \left(\boldsymbol{\Sigma}\left(\underline{d}\,(\underline{r},t), \underline{\dot{d}}\,(\underline{r},t) \right) \underline{n} + \beta \underline{d}\,(\underline{r},t) - \underline{f}_R\,(\underline{r},t) \right) \mathrm{d}\underline{r} = 0\,.
$$

$$\tag{3.16}$$

As this equation must hold for all admissible functions $\underline{v}$, the individual integrals must be equal to zero. Similar to (3.14), this is fulfilled when the object is in equilibrium, which now also depends from the damping forces.

Modeling Uncertainties

As is commonly known, a model can reproduce the behavior of the object only with a limited reliability, since it approximates the motion of the real object by using available noisy data and not exactly known physical information. As a result, the response of the model is corrupted by inaccuracies due to limited precision of the model. The main sources of these inaccuracies, as well as their models, are introduced in this section.

Main Sources of Uncertainties   Generally, the main source of the model inaccuracy is lack of knowledge. For example, the object's structure and behavior are very complex and cannot be resolved at all scales. The main reasons for that is imperfect knowledge of the object's behavior, e.g., unpredictable object's motion, such as extrasystole during beating heart operation, or incomplete data availability. Furthermore, the randomness of the model raises from the input forces corrupted by inaccuracies, as well as from initial conditions corrupted by stochastic perturbations.

Models of Uncertainties and Their Assumptions   As introduced in Section 1.3.2, there are two types of inaccuracies: systematic and stochastic. The systematic inaccuracies represent deterministic errors of model. The stochastic inaccuracies stand for random perturbations. By assuming that the latter are reasonably characterized by additive zero-mean Gaussian random noise, the quantities corrupted by these uncertainties can be sufficiently accurately described by Gaussian random fields, introduced in [216]. The modeling of random quantity by Gaussian field has the advantage of its unique specification by the first two statistical moments, i.e., expectation and covariance. This field is represented by an infinite sequence of continuous Gaussian random variables indexed over the spatial coordinates $\underline{r} \in \overline{\Omega}$ and time $t \in \mathcal{I} \subseteq \mathbb{R}_+$.

It should be noted that the following model inaccuracies, caused by imperfect knowledge of the object's behavior, are assumed to be negligible:

- Unappreciated influences from external environment, e.g., due to simplified embedding of the considered heart area in the surrounding tissues.

- Unpredictable behavior of the object, e.g., the extrasystole.

- Nonlinearities of the heart material.

- Initial stresses and strains of the heart surface.

These uncertainties are difficult to describe mathematically. Their influence can be estimated experimentally by using a highly complex model, intentionally neglecting its levels of sophistication.

Uncertainties of Displacement and Position Fields The displacement field $\underline{d} \in \mathcal{D}$ that satisfies the variational equation (3.16), becomes random due to stochastic uncertainties incorporated in the model. Hence, this field consists of an infinite sequence of random variables

$$\underline{\boldsymbol{d}}\left(\underline{r}, t\right) \sim \mathcal{N}\left(\underline{\mu}^d\left(\underline{r}, t\right), \boldsymbol{\Sigma}^d\left(\underline{r}, t\right)\right), \ \underline{r} \in \overline{\Omega}, \ t \in \mathcal{I} \subseteq \mathbb{R}_+ \qquad (3.17)$$

described by a Gaussian density function $\mathcal{N}$ that is defined by a space-time dependent mean function $\underline{\mu}^d$ and a covariance function $\boldsymbol{\Sigma}^d$. The symbol $\sim$ indicates here the distribution operator. Furthermore, according to (3.2), the position field $\underline{r}^\psi \in \overline{\Omega}^\psi$, approximating the position of the real object, is also random Gaussian. It consists of the sequence of random variables

$$\underline{\boldsymbol{r}}^\psi\left(\underline{r}, t\right) = \underline{r} + \underline{\boldsymbol{d}}\left(\underline{r}, t\right), \ \underline{\boldsymbol{r}}^\psi\left(\underline{r}, t\right) \sim \mathcal{N}\left(\underline{\mu}^{r^\psi}\left(\underline{r}, t\right), \boldsymbol{\Sigma}^{r^\psi}\left(\underline{r}, t\right)\right) \qquad (3.18)$$

with respective mean and covariance functions indexed over the spatial coordinates $\underline{r} \in \overline{\Omega}$ and time $t \in \mathcal{I} \subseteq \mathbb{R}_+$. It should be noted that, as obvious from (3.18), the randomness of the position field is inherited from the randomness of the displacement field.

In this context, it is worth mentioning that the undeformed configuration of the model is assumed to be exactly known. The reason for such an assumption is that the stochastic errors of this configuration will be eliminated in Chapter 4 by averaging the position of the object over an initialization time period. However, the systematic errors of this configuration, such as offset, are difficult to detect. One of the possibilities to solve this problem is to use the set-valued representations [113, 185] of these errors. For this purpose, the information about the nearest neighborhood of every point in the undeformed configuration of the model can be used for constructing the bounded set that describes the possible positions of this point. As a result, according to (3.18), the deformed configuration of the model $\underline{r}^{\psi} \in \overline{\Omega}^{\psi}$ will be defined by sets of densities [147], which can be estimated by set-based filtering approaches [154, 155]. The challenge lies in the mathematical description of the deformed configuration of the model because of the nonlinear dependence of the displacement field on the undeformed configuration of the model.

Uncertainties of Input Forces and Boundary Conditions   There are two reasons for inaccuracies introduced in the model by input forces. First of all, there are systematic errors primarily defined by unknown forces acting on the heart wall. For example, the space-time dependent muscle strengths $\underline{f}_R$ subjecting the Robin boundary Γ_R and body forces $\underline{f}_I$ are unknown. Furthermore, the excitation on the Neumann boundary $\underline{f}_N$ is not exactly known. Although the pressure at all points on Neumann boundary Γ_N is nonuniform due to the fact that the cardiac contractility is caused by inhomogeneous muscle strengths, only uniform pressure can be measured inside the cardiac chamber. With the aim of considering also the nonuniform excitation, the pressure force acting on the Neumann boundary is decomposed on measured uniform and unknown nonuniform parts according to

$$\underline{f}_N\left(\underline{r}, t\right) = \underline{f}_N^u\left(\underline{r}, t\right) + \underline{f}_N^n\left(\underline{r}, t\right).\tag{3.19}$$

The nonuniform part $\underline{f}_N^n$ is defined not only by unknown nonuniform pressure inside the heart chamber, but also by unknown systematic uncertainties of the measured uniform pressure signal, such as, e.g., drift and offset occurring due to inaccurate calibration of the heart catheter.

The second reason for modeling inaccuracies caused by input forces, are stochastic uncertainties, which introduce the random errors. Hence, the uncertainties of the uniform part of the pressure measured by a heart catheter, are caused by the measurement noise, which is assumed to be additive zero-mean Gaussian. Therefore, similarly to the displacement field, the uniform part of the pressure force is defined by the Gaussian random field consisting of the infinite sequence of Gaussian random variables

$$\underline{f}_N^u\left(\underline{r},t\right) \sim \mathcal{N}\left(\underline{\mu}^{f_N^u}\left(\underline{r},t\right), \boldsymbol{\Sigma}^{f_N^u}\left(\underline{r},t\right)\right), \ \underline{r}\in\overline{\Omega}, \ t\in\mathcal{I}\subseteq\mathbb{R}_+ \qquad (3.20)$$

with mean and covariance defined by respective functions.

Uncertainties of Initial Conditions In addition, the initial conditions introduce inaccuracies in the physical model. Since the heart is continuously moving, information about the position of the heart wall at initial time step must be extracted from measurements provided by a camera system. However, these measurements are corrupted by stochastic perturbations caused, e.g., by noise of the feature extraction or changing light conditions, and assumed to be additive Gaussian and zero-mean. As a result, the uncertainties of the initial conditions are inherited from the noise of the camera system measurements. This allows us to characterize the initial values of the displacement and velocity fields by Gaussian random fields represented by infinite sequences of Gaussian random variables

$$\begin{aligned} \underline{d}_0\left(\underline{r}\right) &\sim \mathcal{N}\left(\underline{\mu}_0^d\left(\underline{r}\right), \boldsymbol{\Sigma}_0^d\left(\underline{r}\right)\right), \ \underline{r}\in\overline{\Omega}, \\ \underline{v}_0\left(\underline{r}\right) &\sim \mathcal{N}\left(\underline{\mu}_0^v\left(\underline{r}\right), \boldsymbol{\Sigma}_0^v\left(\underline{r}\right)\right), \ \underline{r}\in\overline{\Omega}, \end{aligned} \qquad (3.21)$$

which mean and covariance are defined by respective space-dependent functions.

Physical Model in Continuous Form

As a result, by extending (3.16) by the models of inaccuracies mathematically described by (3.17), (3.19), (3.20), and (3.21), the probabilistic model approximating the heart wall behavior is formulated. It is represented by the boundary and initial value problem in form of the system of hyperbolic stochastic partial differential equations

$$\mathbb{A}\left[\underline{d}\left(\underline{r},t\right)\right]:=\rho\underline{\ddot{d}}\left(\underline{r},t\right)+\rho\kappa\underline{\dot{d}}\left(\underline{r},t\right)-\operatorname{div}\boldsymbol{\Sigma}\left(\underline{d}\left(\underline{r},t\right),\underline{\dot{d}}\left(\underline{r},t\right)\right)=\underline{f}_{I}\left(\underline{r},t\right)\;,\;\;\underline{r}\in\Omega\,,$$

$$\mathbb{B}_N\left[\underline{d}\left(\underline{r},t\right)\right]:=\boldsymbol{\Sigma}\left(\underline{d}\left(\underline{r},t\right),\underline{\dot{d}}\left(\underline{r},t\right)\right)\underline{n}=\underline{f}_N^{u}\left(\underline{r},t\right)+\underline{f}_N^{n}\left(\underline{r},t\right)\;,\;\;\underline{r}\in\Gamma_N\,,$$

$$\mathbb{B}_R\left[\underline{d}\left(\underline{r},t\right)\right]:=\beta\underline{d}\left(\underline{r},t\right)+\boldsymbol{\Sigma}\left(\underline{d}\left(\underline{r},t\right),\underline{\dot{d}}\left(\underline{r},t\right)\right)\underline{n}=\underline{f}_R\left(\underline{r},t\right)\;,\;\;\underline{r}\in\Gamma_R\,,$$

$$\$_D\left[\underline{d}\left(\underline{r},t_0\right)\right]:=\underline{d}\left(\underline{r},t_0\right)=\underline{d}_0\left(\underline{r}\right)\;,\;\;\underline{r}\in\overline{\Omega}\,,$$

$$\$_V\left[\underline{d}\left(\underline{r},t_0\right)\right]:=\underline{\dot{d}}\left(\underline{r},t_0\right)=\underline{v}_0\left(\underline{r}\right)\;,\;\;\underline{r}\in\overline{\Omega}\,,$$

$$(3.22)$$

which are linear. By way of introduction of the concise notation, the left parts of the equations are denoted by functionals, e.g., the displacement field on the Robin boundary Γ_R is described by the functional $\mathbb{B}_R$. On the Neumann boundary Γ_N, this field must satisfy the functional $\mathbb{B}_N$, while the displacement field in the interior Ω is determined by the functional $\mathbb{A}$. The initial conditions of the displacement and velocity fields are introduced by the functionals $\$_D$ and $\$_V$.

In order to solve this problem, a complete description of the random displacement field $\underline{d}$ consisting of the infinite sequence of random variables has to be found. Due to the Gaussian assumption, this field is uniquely characterized by the first two statistical moments of all random variables of this field.

3.3   Physics-Based System Model

The main point of this section is to establish the physics-based system model, which represents the part of the state-space model introduced in its general form in (3.1). The system model describes the propagation of the system state and is aimed at estimating the heart wall motion by means of stochastic filters.

The first important characteristic of the proposed model is the fact that it incorporates the information about the physical properties of the object under consideration. This information is inherited from the system of stochastic partial differential equations, which serves as a starting point for derivation of the model. Another particularity of this physics-based system

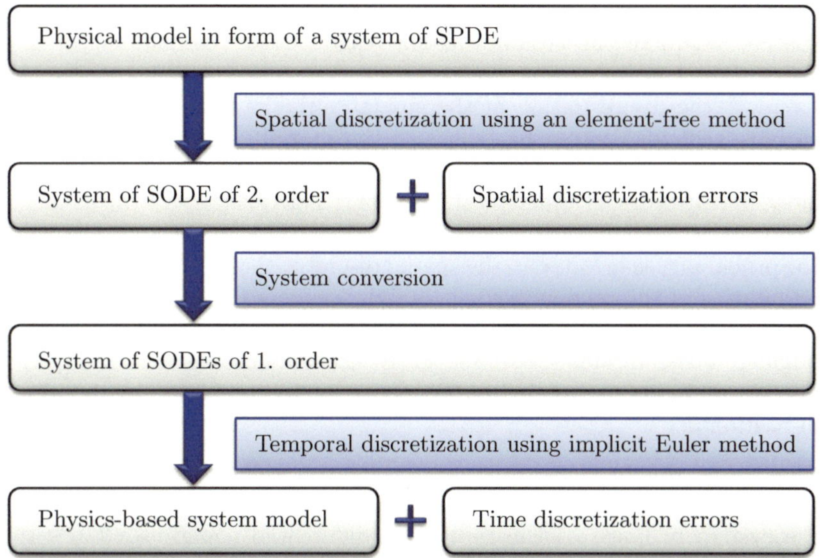

Figure 3.5: Derivation of the state-space model from the physical model using the methods for solution of the system of stochastic partial differential equations (SPDEs). This system is converted in the system of stochastic ordinary differential equations (SODEs) using meshless local Petrov-Galerkin mixed collocation method [14]. For the temporal discretization, the implicit Euler method, e.g., introduced in [38], is applied. The numerical solution of the system of SPDEs introduces additional errors in the system model.

model is that stochastic and systematic modeling errors are considered over all stages of the system model derivation, as depicted in Fig. 3.5. These errors are inherited from the inaccuracies of the physical model introduced in Section 3.2. Furthermore, additional numerical errors arise due to application of numerical methods for derivation of the system model.

For conversion of the system of hyperbolic stochastic partial differential equations (3.22) into a discrete state-space form (3.1), the random fields must be discretized not only in the spatial and temporal domains, as it is usual for deterministic partial differential equations, but also in the probability domain. The key idea of the discretization is inherited from the method of lines [184], which entails considering the dependences from spatial and temporal variables as not equal but as the mapping of spatial variables at every time step. As illustrated in Fig. 3.5, this enables foremost the discretization of the spatial variables, where the temporal variable remains continuous. Through the spatial discretization, the system of stochastic partial differential equations is converted into a system of stochastic ordinary differential equations. For this purpose, an element-free method is applied, as proposed in Section 3.3.1. The discretization of the probability is involved in the spatial discretization and leads to the representation of the random fields by a finite sequence [123] of random variables at every time step. This sequence is finite because it involves a finite number of the variables. Consequently, the implicit Euler method, e.g., introduced in [38], is applied for the temporal discretization, as described in Section 3.3.2.

As a result, due to approximation of the heart wall by the physical body with the linear viscoelastic behavior that is mathematically described by the system of linear stochastic partial differential equations (3.22), the physics-based system model (3.1) is also linear

$$\underline{z}_{k+1} = \mathbf{A}_k \underline{z}_k + \mathbf{B}_k \underline{\hat{u}}_k + \underline{s}_k + \underline{w}_k^z \ . \tag{3.23}$$

It should be noted that physical knowledge about the object under observation is incorporated in the real-valued system $\mathbf{A}_k \in \mathbb{R}^{n_z \times n_z}$ and input $\mathbf{B}_k \in \mathbb{R}^{n_z \times n_{\hat{u}}}$ matrices, which depend on the model parameters with the physical meaning, such as, e.g., material density ρ, or Young modulus E. As a consequence of recursive processing of the system model, these parameters influence the system state at every time step. Hence, the system state $\underline{z}_{k+1} \in \mathbb{R}^{n_z}$ at time step t_{k+1} depends implicitly on the physical properties. Here, the systematic modeling errors are denoted by the vector $\underline{s}_k \in \mathbb{R}^{n_s}$, the stochastic modeling errors are described by the random vector $\underline{w}_k^z \in \mathbb{R}^{n_{w^z}}$. The known input of the system model is represented by vector $\underline{\hat{u}}_k \in \mathbb{R}^{n_{\hat{u}}}$.

3.3.1 Element-Free Spatial Discretization

This section deals with the first stage of the conversion of the physical model formulated by the system of stochastic partial differential equations (3.22) into a physics-based system model (3.23). Since this conversion is based on the methods for a solution of stochastic partial differential equations, first, an overview of these methods is given. Then, the system of stochastic partial differential equations is converted in the system of stochastic ordinary differential equations by exploiting the appropriate element-free method and straightforward discretization of random fields by the finite sequence of random variables.

Overview of Numerical Methods for Stochastic Partial Differential Equations

In this section, an overview of numerical methods for a solution of stochastic partial differential equations is set forth. Since these methods combine probability theory with standard methods for the solution of deterministic partial differential equations, the numerical methods applied only for solution of deterministic problems are also analyzed with respect to their efficiency and accuracy.

Techniques for Solving Stochastic Partial Differential Equations In contrast to deterministic partial differential equations, whose solution is usually represented by the field describing one realization of the system state, stochastic partial differential equations account for randomness of the underlying probabilistic model. Their solution is a random field characterized by statistical moments, i.e., mean, variance, etc., which are described by space and time-continuous random functions.

One possible way to solve the stochastic partial differential equations numerically is to apply a direct method, like a Monte Carlo method [68, 212] or one of its derivatives [98, 153], which implies the solution of one deterministic differential equation, e.g., by classical finite element method, for each realization of the random parameter. The desired expectation of the random field is then approximated by a sample average of independent identically distributed realizations. Although the convergence rate of this method is typically very low, its main advantage is that no assumptions on the small amount of the noise are made.

Another way for solving stochastic partial differential equations is to use the methods that convert the original stochastic problem into the deterministic one by using different methods for discretization of the random fields. These methods, e.g., stochastic finite element methods [6, 7, 17, 18, 221], usually employ a standard finite element approximation in the space domain and, e.g., polynomial or point-based approximation in the probability domain. The main advantage of these methods is a faster convergence rate.

Methods for Discretization of Random Fields   A comprehensive overview of the methods for discretization of the random fields is given in [202, 218]. Generally, these methods can be divided in three groups. The first group consists of point discretization methods, such as, e.g., midpoint method [107], shape function method [130,131], or integration point method [138], which discretize the random field by the set of random variables that represent the values of the random field at some selected points. The second group includes methods that discretize the random field by the random variables averaged over local domains, such as, e.g., the weighted integral method [59,207] or the spatial average method [217]. Finally, the third group consists of series expansion methods, e.g., Karhunen-Loéve expansion or polynomial chaos expansion [196], which approximate the random field by truncated series involving random variables and deterministic functions.

Usually combined with classical methods for spatial discretization, like finite element methods, the methods for discretization of the random fields approximate the random field based on elements. For example, in the midpoint method, the value of random field over an element is represented by its value at the central point of the element. A recent numerical technique, which is still under development, is the use of element-free numerical methods, e.g., an element-free Galerkin method [10,168,169], instead of a classical finite element method. In this technique, the approximation of the random fields may be based upon arbitrary points distributed in the spatial domain.

Classification of Numerical Methods for Partial Differential Equations
Commonly, the numerical methods for the solution of the stochastic, as well as deterministic partial differential equations, can be divided on element-based and element-free methods, as illustrated in Fig. 3.6. In the following section, these groups of methods will be briefly presented. A comprehensive overview of the element-free methods in respect to the solution of the deterministic problems can be found in [13,16,69,126].

1) Element-Based and Element-Free Methods   The group of element-based methods involves, for example, finite element [33], finite difference [44], as well as finite volume [65] methods. The element-free methods

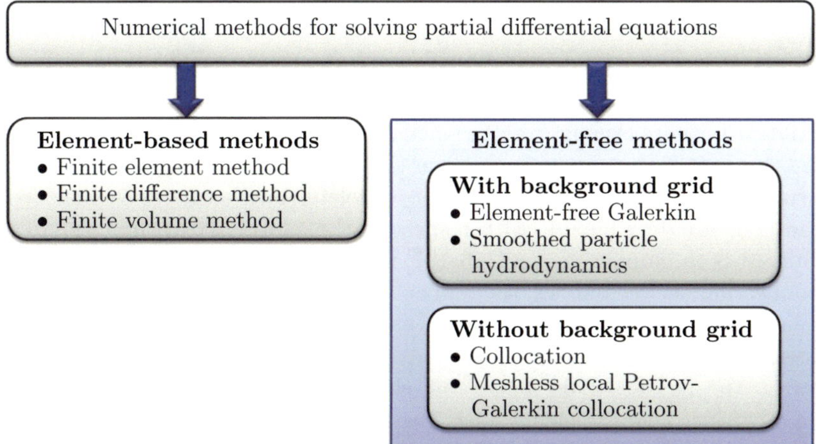

Figure 3.6: Classification of the numerical methods for solving partial differential equations. The element-based methods discretize the spatial domain by elements and approximate the solution of partial differential equations using piecewise polynomial functions defined over these elements. The element-free methods represent the spatial domain by a set of points and make use of the point-based approximation of the solution of the partial differential equation. Some of these methods introduce a background grid to enable a numerical integration.

include such methods as meshless local Petrov-Galerkin mixed collocation [14] or collocation [102].

The main differences between the classical element-based methods and element-free methods are in the discretization of the spatial domain of the partial differential equation and in the approximation of the solution of this equation. The element-based methods discretize the spatial domain by elements and use, e.g., piecewise polynomial functions defined over these elements, such as Legendre or Chebyshev polynomials [103], for approximating the solution of the partial differential equation. It should be noted that vertices of the elements are defined by the discretization points distributed in the spatial domain. Therefore, the connection between these points must be specified for constructing the elements. This information is superfluous for element-free methods. By representing the spatial domain by a set of points, also called model nodes, they make use of point-based approximation of the solution of the partial differential equation, e.g., by radial basic functions [36] or moving least squares functions [126].

In comparison to element-based methods, element-free methods have significant benefits arising from the lack of elements. First of all, the preprocessing is less time-consuming since the conform element mesh, which is crucial for element-based methods, is obsolete. Furthermore, element-free methods are very flexible with respect to the adaptivity of the discretization. It is commonly known [116, 127] that due to change of the predefined connection between nodes, the refinement of the discretization by insertion or removing the nodes can cause severe numerical errors in element-based methods. The element-free methods do not require the connection between the nodes. This makes the changing of their nodes distribution easier. Moreover, the element-free methods are more suitable for handling strong deformations in comparison to element-based methods. The main reason for that is the fact that in case of strong deformations, the element-based methods must cope with a distortion of the elements, which decrease the accuracy of the solution of a partial differential equation.

2) Element-Free Methods with and without Background Grid   It is important to note that the relationship of some methods to the element-free group is too vague. These methods introduce the background grid that is nothing more than a net of elements known from element-based methods. For example, the smoothed particle hydrodynamics method [72, 134, 145, 146] exploiting integral representation of the solution of the partial differential equation, requires a grid for the discretization of the integrals. In addition, the diffuse element method [151] and the element-free Galerkin method [27] make use of a grid-based representation of the solution domain to enable a numerical integration over the spatial domain. Furthermore, the meshless local Petrov-Galerkin method [12] divides the solution domain on subdomains, which are discretized by a grid. The latter three methods are weak form methods. They minimize an average error over the solution domain or, as it is in case of meshless local Petrov-Galerkin method, its subdomains. For this purpose, an integration is required. In contrast, strong form methods, such as collocation methods, e.g., the global collocation [102], the hermite-based symmetric collocation method [67], the Trefftz collocation method [121] or the meshless local Petrov-Galerkin mixed collocation method [14], minimize the approximation errors only on model nodes. It is for this reason that they do not need any elements, not even in form of a background grid.

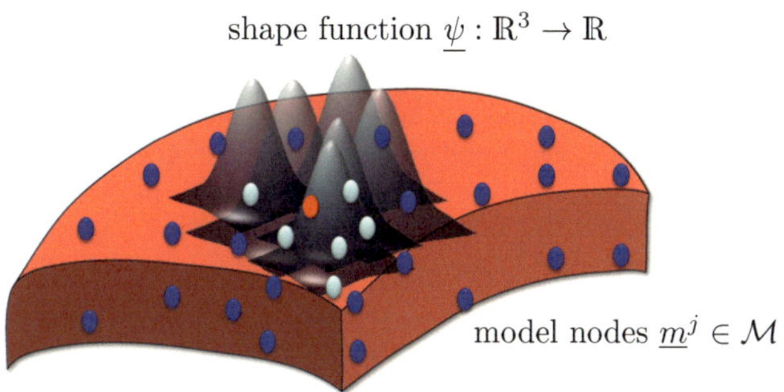

Figure 3.7: Discretization of the spatial domain by a finite set of points $\underline{m}^j \in \mathcal{M}$ and point-based approximation of the displacement field at point $\underline{r}$ by a series of shape functions $\underline{\psi}$ centered at model nodes $\underline{m}^j \in \mathcal{V}$ in the neighborhood of this point.

The element-free methods without background grid have inherent advantages over the methods with background grid, due to complete elimination of the element mesh and absence of integration over the spatial domain. They are easy to implement, highly flexible in the adaptation of the discretization, and they achieve more efficiency.

Spatial Discretization

This section considers the spatial discretization of the system of stochastic partial differential equations. For this purpose, firstly, the spatial domain of the model is discretized by a set of the model nodes. Then, these nodes are used for approximating the displacement field and its derivatives at arbitrary discrete points in the spatial domain. In the next step, for minimizing the approximation errors, an appropriate element-free method is applied. As a result, the system of stochastic partial differential equations is converted in a system of the stochastic ordinary differential equations.

Discretization of the Spatial Domain For spatial discretization, first of all, the spatial domain $\overline{\Omega} \subseteq \mathbb{R}^3$ of the physical model proposed in Section 3.2 is discretized by a finite set $\mathcal{M} := \left\{ \underline{m}^j \right\}_{j=1}^{N_M}$ of points in $\mathbb{R}^3$ with coordinates $\underline{m}^j \in \mathcal{M} \subseteq \overline{\Omega}$, as shown in Fig. 3.7. These points are referred to as model nodes.

It should be emphasized that, in contrast to classical discretization of the spatial domain by elements [33], here, the connectivity between the model nodes is not determined. The reason for this is that the element-free method without background grid will be applied for spatial discretization of the model. Therefore, a priori information about the relationship between the nodes is superfluous, because no elements are needed for the discretization of the system of stochastic partial differential equations.

Approximation of the Displacement Field   As a further step towards the spatially discrete system, the random displacement field representing the solution of the system of stochastic partial differential equations (3.22), as well as the derivatives of this field are approximated by a series of space-dependent shape functions, e.g., moving least squares functions [126] or radial basic functions [36], weighted by random nodal values. With the aim of achieving the local character of this approximation, only the model nodes in the neighborhood of the arbitrary point $\underline{r} = [x^r, y^r, z^r]^{\mathrm{T}}$ are used for approximating the value of the fields at this point, as schematically illustrated in Fig. 3.7. Here, the displacement of the point in red is approximated based on the neighboring points in green. Therefore, when the neighboring model nodes are collected in the set $\mathcal{V} \subseteq \mathcal{M}$, the random displacement field $\underline{d} \in \mathcal{D}$ at every point $\underline{r} \in \overline{\Omega}$ can be approximated by the function $\underline{d}^h : \overline{\Omega} \times \mathcal{V} \times \mathbb{R}_+ \to \mathcal{D}^h$ according to

$$\underline{d}^h(\underline{r}, \underline{m}, t) = \sum_{j \in N_V} \mathbf{\Phi}^j(\underline{r}, \underline{m}^j) \underline{c}^j(t), \qquad (3.24)$$

where N_V denotes the number of model nodes $\underline{m}^j$ in the set $\mathcal{V}$. The coordinates of these nodes are collected in the vector $\underline{m} \in \mathbb{R}^{3N_V}$. Elements on the main diagonal of the matrix function

$$\mathbf{\Phi}^j(\underline{r}, \underline{m}^j) = \mathrm{diag}\left\{\underline{\psi}(\underline{r}, \underline{m}^j)\right\} \in \mathbb{R}^{3 \times 3} \qquad (3.25)$$

are defined by space-dependent shape functions ψ, e.g., moving least squares shape functions [67], collected in the vector $\underline{\psi} \in \mathbb{R}^3$. These functions are centered at every model node $\underline{m}^j \in \mathcal{V}$ and evaluated at every point $\underline{r} \in \overline{\Omega}$. Every jth nodal value is represented by a random vector $\underline{c}^j \in \mathbb{R}^3$, which is time-dependent.

It should be noted that according to (3.24), the spatial and temporal dependences of the random displacement field are separated. This enables the approximation of the space and time derivatives independently from each other. For example, the first and the second time derivatives of the random displacement field approximated by

$$\dot{\underline{d}}^h(\underline{r},\underline{m},t) = \sum_{j\in N_V} \mathbf{\Phi}^j(\underline{r},\underline{m}^j)\dot{\underline{c}}^j(t)\,, \quad \ddot{\underline{d}}^h(\underline{r},\underline{m},t) = \sum_{j\in N_V} \mathbf{\Phi}^j(\underline{r},\underline{m}^j)\ddot{\underline{c}}^j(t)$$

(3.26)

are determined only by the time derivatives of the nodal values. The further meaning of the approximation of the displacement field (3.24) is the projection of the infinite dimensional solution of the system of stochastic partial differential equations (3.22) into a finite dimensional subspace, defined by model nodes and shape functions. This projection is necessary for the minimization of the approximation error that is evaluated in this subspace.

For computation of the matrix of shape functions (3.25), the model nodes within the neighborhood of a point $\underline{r}$ are determined by selecting the nodes near to this point. In elastically deformable objects, the neighborhood remains constant, so that the same shape matrices $\mathbf{\Phi}^j$ are used for the computation of the displacement field (3.26), at every time step. This is in contrast to models of the physical phenomena with changing neighborhood, like gas, fire or water. It is worth mentioning that in this thesis, the neighborhood may change, due to refinement of the spatial discretization by inserting model nodes, as introduced in Chapter 6.

Approximation of Strain and Stress Tensors   The strain and stress tensors involve spatial derivatives of the random displacement field. For approximating these derivatives, first, the strain $\boldsymbol{\mathcal{E}} \in \mathbb{R}^{3\times3}$ and the stress $\boldsymbol{\Sigma} \in \mathbb{R}^{3\times3}$ tensors given in (3.6) and (3.3) are converted in the random strain and stress fields represented by the vectors

$$\underline{\epsilon} = \left[\epsilon_{11}, \epsilon_{22}, \epsilon_{33}, 2\epsilon_{23}, 2\epsilon_{13}, 2\epsilon_{12}\right]^{\mathrm{T}}\,, \quad \underline{\sigma} = \left[\sigma_{11}, \sigma_{22}, \sigma_{33}, 2\sigma_{23}, 2\sigma_{13}, 2\sigma_{12}\right]^{\mathrm{T}}.$$

This compact representation of these tensors is evidenced by the fact that the tensors are symmetric, i.e., shear strains $\epsilon_{12} = \epsilon_{21}$, $\epsilon_{13} = \epsilon_{31}$, $\epsilon_{23} = \epsilon_{32}$, as well as corresponding stresses are equal to each other.

Then, by rewriting (3.4) in the vector form

$$\underline{\epsilon}\left(\underline{d}\left(\underline{r},t\right)\right) := \mathbf{D}\underline{d}\left(\underline{r},t\right)\,,$$

the strain tensor $\mathcal{E}$ represented by a random strain field $\underline{\epsilon}$ is approximated by

$$\underline{\epsilon}^h\left(\underline{d}^h\left(\underline{r},\underline{m},t\right)\right) = \sum_{j\in N_V} \underbrace{\mathbf{D}\mathbf{\Phi}^j\left(\underline{r},\underline{m}^j\right)}_{\mathbf{B}^j\left(\underline{r},\underline{m}^j\right)}\underline{c}^j\left(t\right), \tag{3.27}$$

where the differential operator $\mathbf{D}$ and the matrix $\mathbf{B}^j$ are defined by

$$\mathbf{D} = \begin{bmatrix} \frac{\partial}{\partial x^r} & 0 & 0 \\ 0 & \frac{\partial}{\partial y^r} & 0 \\ 0 & 0 & \frac{\partial}{\partial z^r} \\ \frac{\partial}{\partial y^r} & \frac{\partial}{\partial x^r} & 0 \\ 0 & \frac{\partial}{\partial z^r} & \frac{\partial}{\partial y^r} \\ \frac{\partial}{\partial z^r} & 0 & \frac{\partial}{\partial x^r} \end{bmatrix}, \quad \mathbf{B}^j = \begin{bmatrix} \frac{\partial\psi\left(\underline{r},\underline{m}^j\right)}{\partial x^r} & 0 & 0 \\ 0 & \frac{\partial\psi\left(\underline{r},\underline{m}^j\right)}{\partial y^r} & 0 \\ 0 & 0 & \frac{\partial\psi\left(\underline{r},\underline{m}^j\right)}{\partial z^r} \\ \frac{\partial\psi\left(\underline{r},\underline{m}^j\right)}{\partial y^r} & \frac{\partial\psi\left(\underline{r},\underline{m}^j\right)}{\partial x^r} & 0 \\ 0 & \frac{\partial\psi\left(\underline{r},\underline{m}^j\right)}{\partial z^r} & \frac{\partial\psi\left(\underline{r},\underline{m}^j\right)}{\partial y^r} \\ \frac{\partial\psi\left(\underline{r},\underline{m}^j\right)}{\partial z^r} & 0 & \frac{\partial\psi\left(\underline{r},\underline{m}^j\right)}{\partial x^r} \end{bmatrix}.$$

The matrix $\mathbf{B}^j$ contains the spatial derivatives of the shape function ψ.

Consequently, for approximating the modified Cauchy's stress tensor $\mathbf{\Sigma}$, this tensor (3.15) is rewritten in the vector form

$$\underline{\sigma}\left(\underline{d}\left(\underline{r},t\right),\underline{\dot{d}}\left(\underline{r},t\right)\right) = \mathbf{C}\underline{\epsilon}\left(\underline{d}\left(\underline{r},t\right)\right) + \gamma\mathbf{C}\underline{\dot{\epsilon}}\left(\underline{d}\left(\underline{r},t\right)\right), \tag{3.28}$$

where the time derivative of the strain field is equal to

$$\underline{\dot{\epsilon}}\left(\underline{d}\left(\underline{r},t\right)\right) = \mathbf{D}\underline{\dot{d}}\left(\underline{r},t\right)$$

and the material matrix

$$\mathbf{C} = \frac{\lambda}{\nu}\begin{bmatrix} 1-\nu & \nu & \nu & 0 & 0 & 0 \\ \nu & 1-\nu & \nu & 0 & 0 & 0 \\ \nu & \nu & 1-\nu & 0 & 0 & 0 \\ 0 & 0 & 0 & \left(1-2\nu\right)/2 & 0 & 0 \\ 0 & 0 & 0 & 0 & \left(1-2\nu\right)/2 & 0 \\ 0 & 0 & 0 & 0 & 0 & \left(1-2\nu\right)/2 \end{bmatrix}$$

$$\tag{3.29}$$

incorporates the physical properties of the object under observation defined by the Poisson ratio ν and the first Lamé constant λ. Then, by substituting the approximation of the strain field (3.27) in (3.28), the modified Cauchy's stress tensor represented by a random stress field is approximated by

$$\underline{\sigma}^h \left(\underline{d}^h \left(\underline{r}, \underline{m}, t \right), \dot{\underline{d}}^h \left(\underline{r}, \underline{m}, t \right) \right) = \sum_{j \in N_V} \mathbf{CB}^j \left(\underline{r}, \underline{m}^j \right) \underline{c}^j \left(t \right) \\ + \sum_{j \in N_V} \gamma \mathbf{CB}_j \left(\underline{r}, \underline{m}^j \right) \dot{\underline{c}}^j \left(t \right) . \tag{3.30}$$

It should be noted that for approximating the divergence of the stress tensor (3.13), which is rewritten in a vector form

$$\mathrm{div}\, \underline{\underline{\Sigma}} \left(\underline{d} \left(\underline{r}, t \right), \dot{\underline{d}} \left(\underline{r}, t \right) \right) := \mathbf{D}^{\mathrm{T}} \underline{\sigma} \left(\underline{d} \left(\underline{r}, t \right), \dot{\underline{d}} \left(\underline{r}, t \right) \right) , \tag{3.31}$$

the second spatial derivative of the shape function ψ should be computed. To overcome this, the methodology proposed in [14] is used. For this purpose, the stress field $\underline{\sigma} \in \mathbb{R}^6$ is approximated by a series of weighted shape functions, similarly to the displacement field (3.24). Then, by substituting this approximation in (3.30) and (3.31), the divergence of the stress field results in

$$\mathbf{D}^{\mathrm{T}} \underline{\sigma}^h \left(\underline{d} \left(\underline{r}, \underline{m}, t \right), \dot{\underline{d}} \left(\underline{r}, \underline{m}, t \right) \right) = \\ = \sum_{j \in N_V} \left(\mathbf{B}^j \left(\underline{r}, \underline{m}^j \right) \right)^{\mathrm{T}} \left(\mathbf{S}^j \left(\underline{r}, \underline{m}^j \right) \right)^{-1} \mathbf{CB}^j \left(\underline{r}, \underline{m}^j \right) \underline{c}^j \left(t \right) \\ + \sum_{j \in N_V} \gamma \left(\mathbf{B}^j \left(\underline{r}, \underline{m}^j \right) \right)^{\mathrm{T}} \left(\mathbf{S}^j \left(\underline{r}, \underline{m}^j \right) \right)^{-1} \mathbf{CB}^j \left(\underline{r}, \underline{m}^j \right) \dot{\underline{c}}^j \left(t \right) .$$

The elements on the main diagonal of the matrix

$$\mathbf{S}^j \left(\underline{r}, \underline{m}^j \right) = \mathrm{diag} \left\{ \underline{\psi}(\underline{r}, \underline{m}^j) \right\} \in \mathbb{R}^{6 \times 6}$$

are the shape functions used for approximation of the random stress field $\underline{\sigma}$.

Approximation of the Continuous Model By substituting the approximations of the random displacement field (3.24), its time derivatives (3.26), as well as strain (3.27) and stress (3.30) fields into the system of stochastic partial differential equations (3.22), the problem of finding the displacement of the object under observation given in (3.16) can be reformulated.

Now, the aim is to find the N_V nodal values $\underline{c}^j$ that satisfy the equation

$$\int_\Omega \underline{v}_\Omega^{\mathrm{T}} \left(\mathbb{A} \left[\underline{d}^h \left(\underline{r}, \underline{m}, t \right) \right] - \underline{f}_I \left(\underline{r}, t \right) \right) \mathrm{d}\underline{r}$$

$$+ \int_{\Gamma_N} \underline{v}_{\Gamma_N}^{\mathrm{T}} \left(\mathbb{B}_N \left[\underline{d}^h \left(\underline{r}, \underline{m}, t \right) \right] - \underline{f}_N^u \left(\underline{r}, t \right) - \underline{f}_N^n \left(\underline{r}, t \right) \right) \mathrm{d}\underline{r} \qquad (3.32)$$

$$+ \int_{\Gamma_R} \underline{v}_{\Gamma_R}^{\mathrm{T}} \left(\mathbb{B}_R [\underline{d}^h \left(\underline{r}, \underline{m}, t \right)] - \underline{f}_R \left(\underline{r}, t \right) \right) \mathrm{d}\underline{r} = 0$$

for initial conditions

$$\$_D \left[\underline{d}^h \left(\underline{r}, \underline{m}, t_0 \right) \right] = \underline{d}_0 \left(\underline{r} \right), \underline{r} \in \overline{\Omega},$$

$$\$_V \left[\underline{d}^h \left(\underline{r}, \underline{m}, t_0 \right) \right] = \underline{v}_0 \left(\underline{r} \right), \underline{r} \in \overline{\Omega}$$

and all arbitrary test functions $\underline{v}_\Omega$, $\underline{v}_{\Gamma_N}$ and $\underline{v}_{\Gamma_R}$ at every time step. The functionals $\mathbb{A}$, $\mathbb{B}_N$ and $\mathbb{B}_R$, as well as $\$_D$ and $\$_D$ are defined in (3.22).

Minimization of Approximation Errors Similarly to (3.16), the approximated model (3.32) is justified within given initial conditions and all defined test functions when the terms in parenthesis are equal to zero. However, this is not the case, due to errors involved by the approximation of the displacement field as well as its space and time derivatives. Therefore, for obtaining the numerical solution of a system of stochastic partial differential equations (3.22), these errors should be minimized. For this purpose, numerical methods presented in the previous section, can be applied.

Meshless Local Petrov-Galerkin Mixed Collocation Method For minimization of approximation errors involved in the model (3.32), the meshless local Petrov-Galerkin mixed collocation method [14] is chosen.

As shown in Fig. 3.5, this method refers to the group of element-free methods without background grid that do not use elements. It minimizes approximation errors on selected collocation points $\underline{r}^i \in \overline{\Omega}$ distributed in the undeformed configuration of the physical model. In this way, the integration over the solution domain $\overline{\Omega}$ or its subdomains, which is common for weak form methods, such as the element-free Galerkin method [27] or the meshless local Petrov-Galerkin methods [12], is avoided. This contributes to the efficiency of numerical calculations.

The next advantage of this method is that the spatial variations of the motion are flexibly handled due to local approximation of the solution.

Notwithstanding the above, there is still a problem that the obtained numerical solution may be inaccurate and instabilities may occur in the case of strong spatial variations of the displacement field. The main reason for this is that the information about the motion of the object between the collocation points is not taken into consideration. However, as it is stated in various works [14, 102], this method, as well as other strong form methods [67, 121], provide highly accurate results for smooth functions even for a small number of collocation points.

Therefore, by approximating the smooth displacement field of the heart wall, which has no strong spatial irregularities, this method can provide a sufficient accuracy of the numerical solution and low complexity of numerical computations.

Spatial Discretization of the Continuous Model According to the meshless local Petrov-Galerkin mixed collocation method, the test functions are assumed to be time-independent. Furthermore, they are represented by Dirac delta series

$$\underline{v}_\Omega = \sum_{i=1}^{N_I} \underline{\zeta}^i \delta\left(\underline{r} - \underline{r}^i\right), \ \underline{v}_{\Gamma_N} = \sum_{i=N_I+1}^{N_N} \underline{\zeta}^i \delta\left(\underline{r} - \underline{r}^i\right), \ \underline{v}_{\Gamma_R} = \sum_{i=N_N+1}^{N_R} \underline{\zeta}^i \delta\left(\underline{r} - \underline{r}^i\right)$$

$$(3.33)$$

centered at the collocation points $\underline{r}^i \in \overline{\Omega}$. It should be noted that the coefficients $\underline{\zeta}^i \in \mathbb{R}^3$ are arbitrary.

Plugging the functions (3.33) into (3.32) yields

$$\sum_{i=1}^{N_I} \left(\underline{\zeta}^i\right)^{\mathrm{T}} \left(\mathbb{A}\left[\underline{d}^h\left(\underline{r}^i, \underline{m}, t\right)\right] - \underline{f}_I\left(\underline{r}^i, t\right)\right)$$

$$+ \sum_{i=N_I+1}^{N_N} \left(\underline{\zeta}^i\right)^{\mathrm{T}} \left(\mathbb{B}_N\left[\underline{d}^h\left(\underline{r}^i, \underline{m}, t\right)\right] - \underline{f}_N^u\left(\underline{r}^i, t\right) - \underline{f}_N^n\left(\underline{r}^i, t\right)\right) \quad (3.34)$$

$$+ \sum_{i=N_N+1}^{N_R} \left(\underline{\zeta}^i\right)^{\mathrm{T}} \left(\mathbb{B}_R\left[\underline{d}^h\left(\underline{r}^i, \underline{m}, t\right)\right] - \underline{f}_R\left(\underline{r}^i, t\right)\right) = 0\,.$$

Since the coefficients $\underline{\zeta}^i$ are arbitrary, (3.34) leads to a system of space-discrete algebraic equations of the form

$$\sum_{j=1}^{N_V} \mathbf{M}^{ij} \underline{\ddot{c}}^j\,(t) + \sum_{j=1}^{N_V} \mathbf{V}^{ij} \underline{\dot{c}}^j\,(t) + \sum_{j=1}^{N_V} \mathbf{K}^{ij} \underline{c}^j\,(t) = \underline{f}^i\,(t) \tag{3.35}$$

for every collocation point $\underline{r}^i \in \overline{\Omega}$. The mass $\mathbf{M}^{ij}$, damping $\mathbf{V}^{ij}$ and stiffness $\mathbf{K}^{ij}$ matrices are defined by

$$\mathbf{M}^{ij} = \begin{cases} \rho \boldsymbol{\Phi}^j\left(\underline{r}^i, \underline{m}^j\right) & \text{if } 1 \le i \le N_I\,, \\ 0 & \text{if } I+1 \le i \le N_N\,, \\ 0 & \text{if } N_N+1 \le i \le N_R\,, \end{cases}$$

$$\mathbf{K}^{ij} = \begin{cases} \left(\mathbf{B}^j\left(\underline{r}^i, \underline{m}^j\right)\right)^{\mathrm{T}} \left(\mathbf{S}^j\left(\underline{r}^i, \underline{m}^j\right)\right)^{-1} \mathbf{CB}^j\left(\underline{r}^i, \underline{m}^j\right) & \text{if } 1 \le i \le N_I\,, \\ \mathbf{CB}^j\left(\underline{r}^i, \underline{m}^j\right) \mathbf{N}^i & \text{if } N_I+1 \le i \le N_N\,, \\ \beta \boldsymbol{\Phi}^j\left(\underline{r}^i, \underline{m}^j\right) + \mathbf{CB}^j\left(\underline{r}^i, \underline{m}^j\right) \mathbf{N}^i & \text{if } N_N+1 \le i \le N_R\,, \end{cases}$$

$$\mathbf{V}^{ij} = \begin{cases} \mathbf{U}^{ij}\left(\underline{r}^i, \underline{m}^j\right) & \text{if } 1 \le i \le N_I\,, \\ \gamma \mathbf{CB}^j\left(\underline{r}^i, \underline{m}^j\right) \mathbf{N}^i & \text{if } N_I+1 \le i \le N_N\,, \\ \gamma \mathbf{CB}^j\left(\underline{r}^i, \underline{m}^j\right) \mathbf{N}^i & \text{if } N_N+1 \le i \le N_R\,, \end{cases}$$

$$\tag{3.36}$$

where the matrix $\mathbf{U}^{ij}$ is achieved as follows

$$\mathbf{U}^{ij}\left(\underline{r}^i, \underline{m}^j\right) = \rho \kappa \boldsymbol{\Phi}^j\left(\underline{r}^i, \underline{m}^j\right) + \gamma \left(\mathbf{B}^j\left(\underline{r}^i, \underline{m}^j\right)\right)^{\mathrm{T}} \left(\mathbf{S}^j\left(\underline{r}^i, \underline{m}^j\right)\right)^{-1} \mathbf{CB}^j\left(\underline{r}^i, \underline{m}^j\right)\,.$$

The matrix

$$\mathbf{N}^i = \begin{bmatrix} x^{n^i} & 0 & 0 & 0 & z^{n^i} & y^{n^i} \\ 0 & y^{n^i} & 0 & z^{n^i} & 0 & x^{n^i} \\ 0 & 0 & z^{n^i} & y^{n^i} & x^{n^i} & 0 \end{bmatrix}$$

consists of the elements of the normal vector $\underline{n}^i = \left[x^{n^i}, y^{n^i}, z^{n^i}\right]^{\mathrm{T}}$ at the surface at point $\underline{r}^i$. It should be noted that the boundary $\partial\Omega$ of the surface is assumed to be sufficiently smooth for the existence of the normal. The random vector $\underline{f}^i$ denoting the model input is determined by the function

$$\underline{f}^i\,(t) = \begin{cases} \underline{f}_I\left(\underline{r}^i, t\right)\,, & \text{if } 1 \le i \le N_I\,, \\ \underline{f}_N^u\left(\underline{r}^i, t\right) + \underline{f}_N^n\left(\underline{r}^i, t\right)\,, & \text{if } N_I+1 \le i \le N_N\,, \\ \underline{f}_R\left(\underline{r}^i, t\right)\,, & \text{if } N_N+1 \le i \le N_R\,, \end{cases} \tag{3.37}$$

whereby the random field of the uniform pressure force described by a space-discrete and time-continuous function $\boldsymbol{f}_{N}^{u}$ is now discretized by a finite sequence of time-dependent random variables assigned to the collocation points $\underline{r}^{i}$.

Uncertainties of the Space-Discrete Model

The uncertainties of the space-discrete and time-continuous model (3.35) are primarily inherited from the uncertainties of the system of stochastic partial differential equations (3.22), which is introduced in Section 3.2. Additionally, systematic errors arise from the spatial discretization of the stochastic partial differential equations.

Uncertainties of Input Forces and Boundary Conditions By rewriting the random input (3.37) of space-discrete model as

$$\boldsymbol{f}^{i}(t) = \hat{\underline{u}}^{i}(t) + \underline{u}^{i}(t) + \underline{w}^{u^{i}}(t) \, , \tag{3.38}$$

it should be noted that the value of the random field $\boldsymbol{f}^{i}$ at point $\underline{r}^{i} \in \overline{\Omega}$ is determined by known pressure force $\hat{\underline{u}}^{i}$ corrupted by additional stochastic perturbations $\underline{w}^{u^{i}}$ and unknown systematic errors $\underline{u}^{i}$. The information about the known pressure force $\hat{\underline{u}}^{i}$ is provided by a heart catheter, whereby the unknown part of the model input $\underline{u}^{i}$ is determined by the muscle strengths $\underline{f}_{R}$, density of volume forces $\underline{f}_{I}$, as well as a nonuniform pressure force $\underline{f}_{N}^{n}$ acting on the point $\underline{r}^{i} \in \overline{\Omega}$. Therefore, the known and unknown parts of the model input are defined by

$$\hat{\underline{u}}^{i}(t) = \begin{cases} \underline{0} & \text{if } 1 \leq i \leq N_{I} \, , \\ \underline{\mu}^{f_{N}^{u}}\left(\underline{r}^{i}, t\right) & \text{if } N_{I} + 1 \leq i \leq N_{N} \, , \\ \underline{0} & \text{if } N_{N} + 1 \leq i \leq N_{R} \, , \end{cases}$$

$$\underline{u}^{i}(t) = \begin{cases} \underline{f}_{I}\left(\underline{r}^{i}, t\right) & \text{if } 1 \leq i \leq N_{I} \, , \\ \underline{f}_{N}^{n}\left(\underline{r}^{i}, t\right) & \text{if } N_{I} + 1 \leq i \leq N_{N} \, , \\ \underline{f}_{R}\left(\underline{r}^{i}, t\right) & \text{if } N_{N} + 1 \leq i \leq N_{R} \, , \end{cases} \tag{3.39}$$

where the vector function $\underline{\mu}^{f_{N}^{u}}$ denotes the mean value of the uniform pressure force extracted from the data provided by the heart catheter.

The stochastic uncertainties

$$\underline{\boldsymbol{w}}^{u^i}(t) = \begin{cases} \underline{0} & \text{if } 1 \le i \le N_I\,, \\ \underline{\boldsymbol{w}}^{f_N^u}\left(\underline{r}^i, t\right) & \text{if } N_I + 1 \le i \le N_N\,, \\ \underline{0} & \text{if } N_N + 1 \le i \le N_R\,, \end{cases} \tag{3.40}$$

whose randomness stems from the measurements of the uniform pressure, are inherited from the random input of the continuous model set forth in (3.35). The measurement noise $\underline{\boldsymbol{w}}^{f_N^u}$ is assumed to be zero-mean temporally-white Gaussian, as introduced in Section 3.2. Since the uniform pressure force affects the points lying at the Neumann boundary Γ_N equally, this excitation is spatially invariant, i.e., isotropic.

Uncertainties of Initial Conditions   For modeling the uncertain initial conditions, it should be remembered that for derivation of the space-discrete model, the problem (3.16) of finding the displacement field $\underline{\boldsymbol{d}} \in \mathcal{D}$ of the object is converted into the problem (3.32) of determining the nodal values $\underline{\boldsymbol{c}}^j \in \mathbb{R}^3$ representing the approximation coefficients. Therefore, the initial conditions of the space-discrete model must be formulated in terms of these coefficients.

Their probability distribution is derived based upon the approximation (3.24) of the displacement by a series of weighted space-dependent shape functions. It should be stressed that the motion of the heart wall can be observed by a camera system only at measurement points $\mathcal{L} = \left\{\underline{l}^i\right\}_{i=1}^{N_L}$, $\mathcal{L} \subset \overline{\Omega}$ distributed in the spatial domain $\overline{\Omega}$. Hence, according to (3.21), the initial values of the displacement and velocity fields at these points are denoted by independent Gaussian random variables

$$\begin{aligned} \underline{\boldsymbol{d}}_0^i &\sim \mathcal{N}\left(\underline{\mu}_0^{d^i}, \boldsymbol{\Sigma}_0^{d^i}\right)\,, \quad i \in \{1,\ldots,N_L\}\,, \\ \underline{\boldsymbol{v}}_0^i &\sim \mathcal{N}\left(\underline{\mu}_0^{v^i}, \boldsymbol{\Sigma}_0^{v^i}\right)\,, \quad i \in \{1,\ldots,N_L\}\,. \end{aligned} \tag{3.41}$$

The covariance matrices $\boldsymbol{\Sigma}_0^{d^i} \in \mathbb{R}^{3\times 3}$ and $\boldsymbol{\Sigma}_0^{v^i} \in \mathbb{R}^{3\times 3}$ characterize the uncertainties of these variables. These uncertainties are represented by camera noise, which is assumed to be spatially and temporally white. Then, from (3.24) it follows that the initial values of the approximation coefficients $\underline{\boldsymbol{c}}^j \in \mathbb{R}^3$ collected in the vector $\underline{\boldsymbol{c}}_0 \in \mathbb{R}^{3N_L}$ are defined by Gaussian random vectors

$$\underline{c}_0 = \underline{\mu}_0^c + \underline{w}_0^c \,, \ \ \underline{c}_0 \sim \mathcal{N}\left(\underline{\mu}_0^c, \Sigma_0^c\right) \,,$$

$$\underline{\dot{c}}_0 = \underline{\mu}_0^{\dot{c}} + \underline{w}_0^{\dot{c}} \,, \ \ \underline{\dot{c}}_0 \sim \mathcal{N}\left(\underline{\mu}_0^{\dot{c}}, \Sigma_0^{\dot{c}}\right) \,, \tag{3.42}$$

which are characterized by the mean vectors and covariance matrices computed according to

$$\underline{\mu}_0^c = (\Phi)^{-1}\,\underline{\mu}_0^d \,, \ \ \Sigma_0^c = \left((\Phi)^{-1}\right)^{\mathrm{T}} \Sigma_0^d\,(\Phi)^{-1} \,,$$

$$\underline{\mu}_0^{\dot{c}} = (\Phi)^{-1}\,\underline{\mu}_0^v \,, \ \ \Sigma_0^{\dot{c}} = \left((\Phi)^{-1}\right)^{\mathrm{T}} \Sigma_0^v\,(\Phi)^{-1} \,.$$

In order to ensure the invertibility $(\Phi)^{-1}$ of the matrix Φ, the same number of the model nodes N_M is chosen as the number of the measurement points N_L. The matrix

$$\Phi = \begin{bmatrix} \Phi^{11} & \Phi^{12} & \cdots & \Phi^{1N_M} \\ \vdots & \vdots & \ddots & \vdots \\ \Phi^{N_L 1} & \Phi^{N_L 2} & \cdots & \Phi^{N_L N_M} \end{bmatrix} \tag{3.43}$$

is assembled from the matrices

$$\Phi^{ij} = \Phi^j\left(\underline{l}^i, \underline{m}^j\right) = \mathrm{diag}\left\{\underline{\psi}\left(\underline{l}^i, \underline{m}^j\right)\right\} \in \mathbb{R}^{3\times 3} \tag{3.44}$$

determined by the matrix function Φ^j, which is defined in (3.25). The elements of the vector $\underline{\psi} \in \mathbb{R}^3$ are values of the space-dependent shape function ψ centered at model node $\underline{m}^j \in \mathcal{M}$ and evaluated at the point $\underline{l}^i \in \mathcal{L}$. Furthermore, the vectors $\underline{\mu}_0^d$ and $\underline{\mu}_0^v$ collect the mean values

$$\underline{\mu}_0^d = \begin{bmatrix} \underline{\mu}_0^{d^1} \\ \underline{\mu}_0^{d^2} \\ \vdots \\ \underline{\mu}_0^{d^{N_L}} \end{bmatrix} \,, \ \ \underline{\mu}_0^v = \begin{bmatrix} \underline{\mu}_0^{v^1} \\ \underline{\mu}_0^{v^2} \\ \vdots \\ \underline{\mu}_0^{v^{N_L}} \end{bmatrix}$$

of the N_L random variables introduced in (3.41). The block-diagonal matrices Σ_0^d and Σ_0^v are assembled from covariance matrices of these variables according to

$$\Sigma_0^d = \mathrm{diag}\left\{\Sigma_0^{d^1}, \Sigma_0^{d^2}, \dots, \Sigma_0^{d^{N_L}}\right\} \,,$$

$$\Sigma_0^v = \mathrm{diag}\left\{\Sigma_0^{v^1}, \Sigma_0^{v^2}, \dots, \Sigma_0^{v^{N_L}}\right\} \,.$$

Spatial Discretization Errors In addition to the uncertainties of the model input and initial conditions inherited from the continuous model, the space-discrete model involves discretization, also known as approximation errors, which arise from the spatial discretization. These errors are systematic. They are defined as the difference between the exact solution of the stochastic partial differential equations system (3.22) and its numerical approximation (3.24) at every time step. Unfortunately, the exact solution of the system of stochastic partial differential equations (3.22) is unavailable. This is the reason that the errors of the spatial discretization can only be estimated. The second important point is that, to our knowledge, there is still no mathematical analysis on the a priori error estimation of the meshless local Petrov-Galerkin mixed collocation method. The main reason for this is that this method does not possess the property of orthogonality between the solution space and error space that builds a basis for the error analysis of the standard finite element method. Therefore, the theory of the finite element method cannot be used for this purpose. In this thesis, the approximation errors are continuously reduced based on the a posteriori error evaluation provided in Chapter 6 that renders the a priori estimation of these errors superfluous.

Resulting Space-Discrete Time-Continuous Model

As a result of spatial discretization, the system of stochastic partial differential equations (3.22) is converted in the system of stochastic ordinary differential equations written in a concise vector-matrix form

$$
\begin{aligned}
\mathbf{M}\ddot{\underline{c}}(t) + \mathbf{V}\dot{\underline{c}}(t) + \mathbf{K}\underline{c}(t) &= \hat{\underline{u}}(t) + \underline{u}(t) + \underline{w}^u(t) \\
\underline{c}(t_0) &= \underline{\mu}_0^c + \underline{w}_0^c, \\
\dot{\underline{c}}(t_0) &= \underline{\mu}_0^{\dot{c}} + \underline{w}_0^{\dot{c}}.
\end{aligned}
\tag{3.45}
$$

Here, the vectors of the known $\hat{\underline{u}} \in \mathbb{R}^{3N_R}$ and unknown $\underline{u} \in \mathbb{R}^{3N_R}$ model input, as well as the vector of the stochastic perturbations $\underline{w}^u \in \mathbb{R}^{3N_R}$ collect values of the respective functions shown in (3.39) and (3.40), which are evaluated at collocation points $\underline{r}^i \in \overline{\Omega}$. The global mass matrix $\mathbf{M} \in \mathbb{R}^{3N_R \times 3N_M}$, damping matrix $\mathbf{V} \in \mathbb{R}^{3N_R \times 3N_M}$ and stiffness matrix $\mathbf{K} \in \mathbb{R}^{3N_R \times 3N_M}$ are assembled from local matrices (3.36) in the same way as the matrix $\mathbf{\Phi} \in \mathbb{R}^{3N_R \times 3N_M}$, which is defined in (3.43).

3.3.2 Temporal Discretization

In this section, the space-discrete physical model, formulated by a system of stochastic ordinary differential equations (3.45), is converted in the space- and time-discrete system model by means of time-discretization. As illustrated in Fig. 3.5, for this purpose, the system of stochastic ordinary differential equations, which is of second order, is first converted into the first order system. Then, the integral equation defining the solution of the latter system is approximated by an implicit Euler method under the assumption of the piecewise constant behavior of system input. As a result, this section concludes with the formulation of the discrete physics-based system model and its uncertainties.

Conversion to a First Order System   The system of stochastic ordinary differential equations (3.45), which is of second order, can be rewritten as a first order time-continuous system

$$\dot{\underline{z}}(t) = \mathbf{A}\underline{z}(t) + \mathbf{B}\left(\hat{\underline{u}}(t) + \underline{u}(t)\right) + \underline{w}^{u}(t) \tag{3.46}$$

characterized by the system state

$$\underline{z}(t) = \begin{bmatrix} \underline{c}(t) \\ \dot{\underline{c}}(t) \end{bmatrix} \tag{3.47}$$

collecting the nodal values and their time-derivatives. The system $\mathbf{A} \in \mathbb{R}^{6N_R \times 6N_M}$ and input $\mathbf{B} \in \mathbb{R}^{6N_R \times 3N_R}$ matrices are defined by

$$\mathbf{A} = \left(\begin{bmatrix} \mathbf{V} & \mathbf{M} \\ \mathbf{I} & \mathbf{O} \end{bmatrix}\right)^{-1} \begin{bmatrix} -\mathbf{K} & \mathbf{O} \\ \mathbf{O} & \mathbf{I} \end{bmatrix},$$

$$\mathbf{B} = \left(\begin{bmatrix} \mathbf{V} & \mathbf{M} \\ \mathbf{I} & \mathbf{O} \end{bmatrix}\right)^{-1} \begin{bmatrix} \mathbf{I} \\ \mathbf{O} \end{bmatrix},$$

where the matrices $\mathbf{M} \in \mathbb{R}^{3N_R \times 3N_M}$, $\mathbf{V} \in \mathbb{R}^{3N_R \times 3N_M}$ and $\mathbf{K} \in \mathbb{R}^{3N_R \times 3N_M}$ stem from (3.45). The matrix $\mathbf{I} \in \mathbb{R}^{3N_R \times 3N_M}$ denotes the identity matrix and the matrix $\mathbf{O} \in \mathbb{R}^{3N_R \times 3N_M}$ stands for zero matrix.

It should be noted that in (3.46), the first time derivatives of nodal values are considered as separate variables.

Numerical Time Integration   An implicit Euler time integration method [75] is applied for temporal discretization of (3.46). In contrast to the explicit methods, such as, e.g., explicit Euler method [75], which are only stable when the step size is small enough, this method has no restrictions on the step size. Of course, the accuracy is lost when the step size becomes large, however, this method will never lose stability [75]. In comparison to the higher order approximation methods, such as implicit Runge-Kutta method, the implicit Euler method is computationally less expensive, but provides less accuracy when the step size is large.

The time-continuous solution of the system (3.46)

$$\underline{z}(t) = e^{\mathbf{A}(t-t_0)}\underline{z}(t_0) + \int_{t_0}^{t} e^{\mathbf{A}(t-\tau)}\mathbf{B}(\tau)\underline{\hat{u}}(\tau)\,\mathrm{d}\tau$$

$$+ \int_{t_0}^{t} e^{\mathbf{A}(t-\tau)}\mathbf{B}(\tau)\left(\underline{u}(\tau) + \underline{w}^{u}(\tau)\right)\mathrm{d}\tau$$

is approximated between two time steps $t = t_k$ and $t_0 = t_{k+1}$ by a rectangular integration

$$\underline{z}(t_{k+1}) \approx \underline{z}(t_k) + \int_{t_k}^{t_{k+1}} \left(\mathbf{A}(\tau)\underline{z}(t_{k+1}) + \mathbf{B}(\tau)(\underline{\hat{u}}(\tau) + \underline{u}(\tau)) + \mathbf{B}(\tau)\underline{w}^{u}(\tau)\right)\mathrm{d}\tau,$$

$$(3.48)$$

where it is assumed that the step size $\Delta t = t_k - t_{k+1}$ is sufficiently small for achieving a high accuracy. In this case, the functions defining the known and unknown model input $\underline{\hat{u}} \in \mathbb{R}^{3N_R}$, $\underline{u} \in \mathbb{R}^{3N_R}$ can be assumed to be piecewise constant between the time steps. Furthermore, the matrices $\mathbf{A}$ and $\mathbf{B}$, which involve physical properties of the object that may vary over time, are supposed to remain constant between the time steps.

From (3.48) it follows that the approximation of the system state at time step $t = t_{k+1}$ is defined by the time-discrete equation

$$\underline{z}_{k+1} = \mathbf{A}_k\underline{z}_k + \mathbf{B}_k\left(\underline{\hat{u}}_k + \underline{u}_k\right) + \underline{w}_k^{z}\,, \qquad (3.49)$$

where the time-discrete system and input matrices denoted by $\mathbf{A}_k \in \mathbb{R}^{6N_R \times 6N_M}$ and $\mathbf{B}_k \in \mathbb{R}^{6N_R \times 3N_R}$ respectively so as the process noise

$\underline{w}_k^z \in \mathbb{R}^{6N_R}$ are approximated by

$$\mathbf{A}_k = \mathbf{I}\left(\mathbf{I} - \int_{t_k}^{t_{k+1}} \mathbf{A}(\tau)\,\mathrm{d}\tau\right)^{-1} \approx \mathbf{I}\left(\mathbf{I} - \mathbf{A}\Delta t\right)^{-1},$$

$$\mathbf{B}_k = \int_{t_k}^{t_{k+1}} \mathbf{B}(\tau)\left(\mathbf{I} - \int_{t_k}^{t_{k+1}} \mathbf{A}(\tau)\mathrm{d}\tau\right)^{-1}\mathrm{d}\tau \approx \mathbf{B}\Delta t\left(\mathbf{I} - \mathbf{A}\Delta t\right)^{-1}, \quad (3.50)$$

$$\underline{w}_k^z = \int_{t_k}^{t_{k+1}} \mathbf{B}(\tau)\left(\mathbf{I} - \int_{t_k}^{t_{k+1}} \mathbf{A}(\tau)\,\mathrm{d}\tau\right)^{-1}\underline{w}^u(\tau)\,\mathrm{d}\tau.$$

As a result of the first two equations (3.50), the system and input matrices of the space- and time-discrete system (3.49) can be written in the concise form

$$\mathbf{A}_k = \left(\begin{bmatrix} \mathbf{V} + \Delta t\mathbf{K} & \mathbf{M} \\ \mathbf{I} & -\mathbf{I}\Delta t \end{bmatrix}\right)^{-1}\begin{bmatrix} \mathbf{V} & \mathbf{M} \\ \mathbf{I} & \mathbf{O} \end{bmatrix},$$

$$\mathbf{B}_k = \left(\begin{bmatrix} \mathbf{V} + \Delta t\mathbf{K} & \mathbf{M} \\ \mathbf{I} & -\mathbf{I}\Delta t \end{bmatrix}\right)^{-1}\begin{bmatrix} \Delta t\mathbf{I} \\ \mathbf{O} \end{bmatrix}. \quad (3.51)$$

It is worth noting that when the number of the model nodes $\underline{m}^j \in \mathcal{M}$ is equal to the number of the collocation points $\underline{r}^i \in \overline{\Omega}$, the inverses of the matrices exist.

Uncertainties of Space- and Time-Discrete Model

The uncertainties of the space- and time-discrete model (3.49) representing the physics-based system model are inherited, first of all, from the space-discrete and time-continuous model (3.45), which is used as a starting point for temporal discretization. Furthermore, since the approximation of the system state (3.49) obtained using implicit Euler method satisfies the continuous equations (3.46) only approximatively, additional systematic errors arise from the time-discretization.

Uncertainties of Model Input   According to (3.50), the stochastic uncertainties of the system model $\underline{w}_k^z$ are determined by the uncertainties $\underline{w}^u$ of the input of space-discrete and time-continuous model (3.45). By assuming that the noise process is stationary, i.e., its probability distribution does not change when shifted in time, the covariance of the process noise can be approximated similarly to [195] by

$$
\mathbf{\Sigma}_k^{w^z} = \int_{t_k}^{t_{k+1}} \left(\mathbf{I} - \int_{t_k}^{t_{k+1}} \mathbf{A}(\tau)\,\mathrm{d}\tau \right)^{-1} \mathbf{B}(\tau) \mathbf{\Sigma}^u(\tau) \mathbf{B}^\mathrm{T}(\tau)
$$

$$
\cdot \left(\left(\mathbf{I} - \int_{t_k}^{t_{k+1}} \mathbf{A}(\tau)\,\mathrm{d}\tau \right)^{-1} \right)^\mathrm{T} \mathrm{d}\tau \approx \mathbf{B}_k \mathbf{\Sigma}_k^u \mathbf{B}_k^\mathrm{T} \, ,
$$

where the covariance matrix of the time-discrete input noise is equal to $\mathbf{\Sigma}_k^u = \frac{\mathbf{\Sigma}^u}{\Delta t}$.

Uncertainties of Initial Conditions   The uncertainties of the initial conditions come directly from the space-discrete model (3.45). In order to represent the initial conditions of the system model in a concise vector-matrix form, the mean values of the approximation coefficients at initial time step t_0 and its derivatives are collected in the vector $\underline{\mu}_0^z = \left[\left(\underline{\mu}_0^c \right)^\mathrm{T}, \left(\underline{\mu}_0^{\dot c} \right)^\mathrm{T} \right]^\mathrm{T}$. Then, the initial conditions can be written in the form

$$
\underline{z}_0 = \underline{\mu}_0^z + \underline{w}_0^z \, ,
$$

where the vector $\underline{w}_0^z = \left[\left(\underline{w}_0^c \right)^\mathrm{T}, \left(\underline{w}_0^{\dot c} \right)^\mathrm{T} \right]^\mathrm{T}$ contains the stochastic errors.

Time Discretization Errors   Before estimating the systematic errors caused by time-discretization, it should be noted that the system state $\underline{z}\,(t_{k+1}) \sim \mathcal{N}\left(\underline{\mu}^z\,(t_{k+1}), \mathbf{\Sigma}^z\,(t_{k+1}) \right)$ that represents the exact solution of the system of stochastic partial differential equations (3.46) at time step t_{k+1} and its approximation $\underline{z}_{k+1} \sim \mathcal{N}\left(\underline{\mu}_{k+1}^z, \mathbf{\Sigma}_{k+1}^z \right)$ are Gaussian distributed with respective mean and covariance functions. Then, the error caused by time-discretization $\underline{\tau}_{k+1} = \underline{\mu}^z\,(t_{k+1}) - \underline{\mu}_{k+1}^z$ is defined as the difference between the mean values of the exact and approximated solutions. According to (3.48), it arises from approximating the derivative of the system state by a rectangular integration

$$
\underline{\dot\mu}^z(t_{k+1}) \approx \lim_{\Delta t \to 0} \frac{1}{\Delta t} \left(\underline{\mu}^z(t_{k+1}) - \underline{\mu}^z(t_k) \right) \, . \tag{3.52}
$$

Therefore, the error $\underline{\tau}_{k+1}$ is equal to a local truncation error of the implicit Euler method, which can be obtained by expanding both parts of (3.52) by the Taylor series around the time step t_k according to

$$
\begin{aligned}
\underline{\tau}_{k+1} &= \frac{1}{\Delta t}\left(\underline{\mu}^z\left(t_{k+1}\right) - \underline{\mu}^z\left(t_k\right)\right) - \underline{\dot{\mu}}^z\left(t_{k+1}\right) \\
&= \underline{\dot{\mu}}^z\left(t_k\right) + \frac{\Delta t}{2}\underline{\ddot{\mu}}^z\left(t_k\right) - \underbrace{\underline{\dot{\mu}}^z\left(t_{k+1}\right)}_{\underline{\dot{\mu}}^z(t_k)+\Delta t\underline{\ddot{\mu}}^z(t_k)} \\
&= -\frac{1}{2}\Delta t\underline{\ddot{\mu}}^z\left(t_k\right) + \mathcal{O}\left(\Delta t^2\right).
\end{aligned}
\tag{3.53}
$$

It should be noted that the function μ^z is twice differentiable with respect to time that follows from the assumption on twice differentiability of the displacement field in respect to time.

As a result, the systematic error caused by the time-discretization is proportional to the size of the time step.

Discrete Physics-Based System Model

To summarize the results, the discrete system model, given in its general form in (3.1), is defined by the discrete function

$$
\underline{z}_{k+1} = \mathbf{A}_k\underline{z}_k + \mathbf{B}_k\underline{\hat{u}}_k + \underline{s}_k + \underline{w}_k^z,
\tag{3.54}
$$

where the systematic errors $\underline{s}_k$ and initial conditions for system state $\underline{z}_0$ are equal to

$$
\underline{s}_k = \mathbf{B}_k\underline{u}_k + \underline{\tau}_k, \quad \underline{z}_0 = \underline{\mu}_0^z + \underline{w}_0^z.
$$

The system noise $\underline{w}_k^z$ is stationary, zero-mean and Gaussian $\underline{w}_k^z \sim \mathcal{N}\left(\underline{0}, \mathbf{\Sigma}_k^{w^z}\right)$ with covariance $\mathbf{\Sigma}_k^{w^z}$.

Derived from the system of stochastic partial differential equations (3.22) by means of space- and time-discretization, the system model (3.54) is physics-based, since its mathematical formulation involves the physical properties of the object under observation. Accordingly, the parameter of the physical model, such as the Young modulus or the Poisson ratio, which define the material properties of the object, determine the system matrices $\mathbf{A}_k$ and $\mathbf{B}_k$ and therefore, also the system state $\underline{z}_{k+1}$ at the next time step.

3.4 Physics-Based Measurement Model

The measurement model, which is introduced in its general form by the measurement equation (3.1), defines the relationship between the system

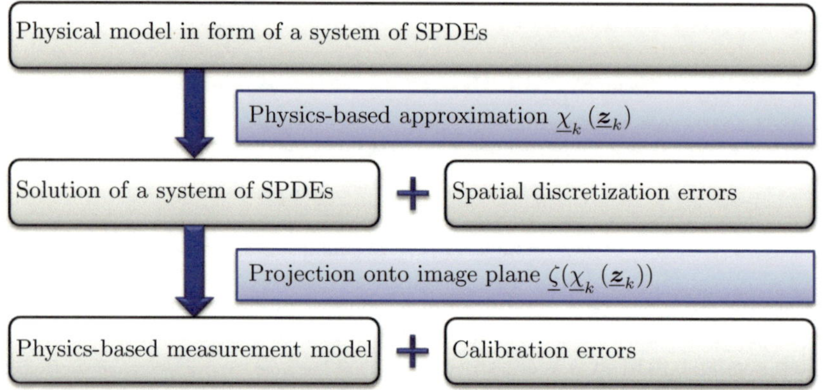

Figure 3.8: Physics-based measurement model is derived from the physical model formulated by a system of stochastic partial differential equations (SPDE). It implicates the numerical solution of this system.

state $\underline{\boldsymbol{z}}_k \in \mathbb{R}^{n_z}$ and measurements $\underline{\hat{y}}_k \in \mathbb{R}^{n_y}$ provided by a trinocular camera system at the current time step t_k. The aim of this section is to derive the measurement model from the underlying physical model that was mathematically formulated in Section 3.2 by a system of stochastic partial differential equations (3.22).

The measurement model is based on the fact that the camera image represents the projection of the object under observation onto the image plane of the camera. Therefore, as illustrated in Fig. 3.8, the deduction of this model can be divided into two steps, leading to the implication of the physical characteristics of the object in the measurement equation. In order to clarify this context, it should be noted that the system state, provided by a physics-based system model, depends on the physical properties of the object, such as, e.g., Poisson ratio. In the first step, this state is used for approximating the current position of the object under observation, which is described by a solution of the system of stochastic partial differential equations. This solution is obtained by a physics-based approximation, defined by the time-dependent function

$$\underline{\chi}_k : \mathbb{R}^{n_z} \to \mathbb{R}^3 \qquad (3.55)$$

that maps the current system state $\underline{\boldsymbol{z}}_k \in \mathbb{R}^{n_z}$ to the current position of the object in $\mathbb{R}^3$. In the next step, this object is projected onto the image plane of the camera. This is described by the function

$$\underline{\zeta} : \mathbb{R}^3 \to \mathbb{R}^2 \tag{3.56}$$

that maps the current position of the object in $\mathbb{R}^3$ to its image projection in $\mathbb{R}^2$.

With the aim of providing an accurate description of the relationship between the system state and measurements, systematic and stochastic errors are considered at every step of the measurement model deduction. Thus, approximation errors are introduced in the measurement model because of the numerical approximation of the solution of the system of stochastic partial differential equations, which is not analytically solvable. Furthermore, calibration inaccuracies contaminate the projection of the approximated position of the object under observation onto the image plane of the camera.

3.4.1 Model of Camera System Measurements

Against the background that the measurement model defines the relationship between the system state and the camera measurements, this section starts with the mathematical description of the measurements. Then, the physics-based measurement model is derived in two steps: physics-based approximation of the object's position and its projection onto the image plane of the camera.

Mathematical Description of Camera Measurements

The motion of the heart surface is observed by a stationary trinocular camera system, which provides at every discrete time step three camera images, as depicted in Fig. 3.9. These images illustrate the position of the heart surface projected onto the image planes of the cameras at respective time steps.

Assumptions With the aim of providing a mathematical description of this projection, the functionality of every camera enumerated by index $j = 1, \ldots, 3$ is approximated by a camera model, which is used by a calibration algorithm for estimating the extrinsic $\mathbf{R}^j$ and intrinsic $\mathbf{K}^j$ camera parameters, as well as translation vector $\underline{t}^j$. Each of the cameras is supposed to be modeled well by a pinhole camera model. The detailed description of this model can be found in [109]. The pinhole camera model

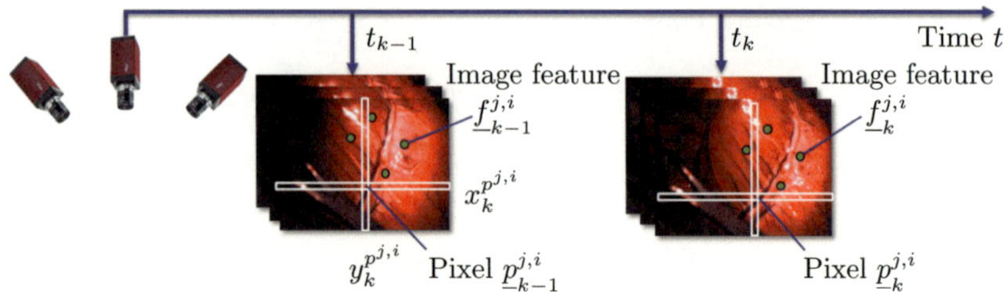

Figure 3.9: Measurement information provided by camera images.

assumes that the three-dimensional points on the visible surface of the object are projected onto the images of every camera according to the central projection. Another assumption made is that the image distortion is negligible. This is the case when the finest professional photographic lenses are used and the cameras are located sufficiently far away from the object under observation. The camera system can be installed for surgical operations on the open beating heart in this way. In case of minimal-invasive operations employing endoscopes, the image distortion may be strong and therefore, must be removed, e.g., by the algorithm proposed in [15].

Image as Set of Pixels The images provided by a trinocular camera system arrive continuously in real time. Each image is represented by a set of pixels with assigned color information. For example, as illustrated in Fig. 3.9, the image provided by camera j at the current time step t_k is defined by a set $\mathcal{P}^j_k = \left\{ \underline{p}^{j,i}_k \right\}^{N_{P^j_k}}_{i=1}$ of pixels $\underline{p}^{j,i}_k := \left[x^{p^{j,i}}_k, y^{p^{j,i}}_k \right]^{\mathrm{T}} \in \mathcal{P}^j_k$ identified by row $x^{p^{j,i}}_k$ and column $y^{p^{j,i}}_k$ indices that are bounded by constant limits of the image resolution.

The image sequence collecting all images provided by camera j over an open time interval $I := [t_1, t_1+T)$, where $T > 0$ and t_1 is initial time step, is described by an unbounded sequence of sets $\mathcal{P}^j_{1:k} = \left\{ \mathcal{P}^j_1, \mathcal{P}^j_2, \ldots, \mathcal{P}^j_{k-1}, \mathcal{P}^j_k \right\}$ enumerated by index $k \in I$. Finally, all three image sequences of the trinocular camera system are denoted by a union of unbounded sequences

$$\mathcal{P}_{1:k} = \bigcup_{j=1}^{n} \mathcal{P}^j_{1:k}, \ n \in \{1, 2, 3\}.$$

Camera Measurements   The common way to reconstruct the position of the object under observation from camera images is to employ a triangulation method introduced in [80]. Unfortunately, in this way, the entire heart surface cannot be reconstructed without further effort. The main reason is that images are represented by discrete sets of pixels. Therefore, when the correspondences between all pixels of the images acquired at the same time step are exactly known, the position of the heart surface can still be obtained only at discrete points. Since the exact correspondences between the pixels are hardly determinable, it is more practicable to look for correspondences between only some image features that can be extracted by proper feature detection algorithms, an overview of which is illustrated in [206].

The positions of all image features extracted from the image of camera j at time step t_k are collected in the set $\mathcal{F}_k^j = \left\{ \underline{f}_k^{j,i} \right\}_{i=1}^{N_{F_k^j}} \subseteq \mathcal{P}_k^j$. Similarly to the pixels, each image feature $\underline{f}_k^{j,i} := \left[x_k^{f^{j,i}}, y_k^{f^{j,i}} \right]^{\mathrm{T}}$ is identified by row $x_k^{f^{j,i}}$ and column $y_k^{f^{j,i}}$ indices. Only image features that have unique correspondences in all three camera images acquired at the same time step represent the camera measurements $\underline{\hat{y}}_k^{j,i} \in \mathcal{Y}_k^j \subseteq \mathcal{F}_k^j$. The image features extracted from all images of the image sequence gained by camera j over an open time interval I are ordered in the sequence of image feature sets $\mathcal{F}_{1:k}^j = \left\{ \mathcal{F}_1^j, \mathcal{F}_2^j, \dots, \mathcal{F}_{k-1}^j, \mathcal{F}_k^j \right\}$. Finally, the image features extracted from all three image sequences are assembled in a union of sequences $\mathcal{F}_{1:k} = \bigcup_{j=1}^{n} \mathcal{F}_{1:k}^j$, $n \in \{1, 2, 3\}$.

Physics-Based Approximation

For reconstructing the three-dimensional position of the heart surface under observation, a physics-based approximation defined in its general form (3.55) by the function $\underline{\chi}_k$, is employed. This function is derived here in a straightforward manner from the numerical solution (3.24) of the system of stochastic partial differential equations (3.22).

In order to define the relationship between the system state $\underline{z}_k \in \mathbb{R}^{n_z}$ and measurements collected in the vector $\underline{\hat{y}}_k \in \mathbb{R}^{n_y}$, it should be noted that image features represent the projections of the measurement points, i.e., landmarks distributed on the heart surface, onto the image plane of the camera. Since the problem of establishing the correspondences between the image features and landmarks is a subject of the next chapter, these correspondences are assumed to be exactly known in this section.

The positions of all landmarks at time step t_k are collected in the set $\mathcal{L}_k = \{\underline{l}_k^i\}_{i=1}^{N_{L_k}} \subseteq \overline{\Omega}_k^\psi$, where the position of every landmark is described by three-dimensional coordinates $\underline{l}_k^i := \left[x_k^{l^i}, y_k^{l^i}, z_k^{l^i}\right]^\mathrm{T} \in \mathcal{L}_k$. It should be emphasized that the set of the landmarks $\mathcal{L}_k$ is a subset of the bounded domain $\overline{\Omega}_k^\psi$ introduced in Section 3.2.

The current position of the landmark $\underline{l}_k^i$ on the heart surface is not exactly known, since the underlying physical model approximating the behavior of the object under observation is corrupted by uncertainties. The displacement at this point is determined by the numerical solution of the system of stochastic partial differential equations (3.24) evaluated on point $\underline{l}^i \in \overline{\Omega}$. The point $\underline{l}^i$ represents the position of the landmark $\underline{l}_k^i$ in the undeformed configuration of the model introduced in Section 3.2. Since the heart is continuously moving, it is difficult to determine its undeformed configuration. Therefore, the point $\underline{l}^i$ is obtained by averaging the measured positions of the landmark $\underline{l}_k^i$ over a certain time interval $I := [t_1, t_1 + T]$, where t_1 is the initial time step and $T > 0$ is the initialization time. These positions are reconstructed from the corresponding image features detected in all three cameras by a triangulation method introduced in [80] and shortly presented in Appendix A. It should be noted that all positions of the landmarks in the undeformed configuration of the model are collected in the set $\mathcal{L} = \{\underline{l}^i\}_{i=1}^{N_L} \subseteq \overline{\Omega}$ that is a subset of the domain $\overline{\Omega}$.

As illustrated in Fig. 3.10, the uncertain position of every landmark $\underline{l}_k^i$ on the heart surface is determined by a physics-based approximation at every time step t_k. This approximation is defined by the function

$$\underline{l}_k^i(\underline{z}_k) = \underline{\chi}_k(\underline{z}_k) = \underline{l}^i + \underline{d}_k(\underline{z}_k) = \underline{l}^i + \sum_{j \in N_V} \mathbf{H}^{ij} \underline{z}_k^j , \tag{3.57}$$

which takes into account that the displacement of the object (3.24) represents the solution of the system of stochastic partial differential equations (3.22). The matrix

$$\mathbf{H}^{ij} = \left[\mathbf{\Phi}^{ij}, \, \mathbf{0}\right] \in \mathbb{R}^{3 \times 6} \tag{3.58}$$

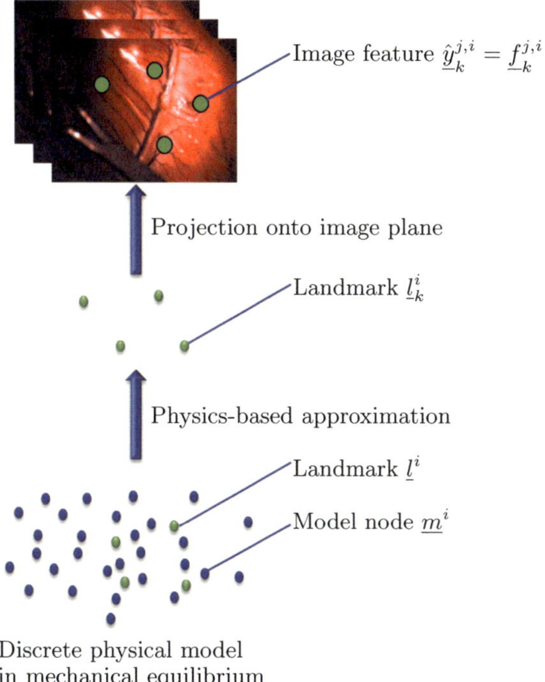

Figure 3.10: Derivation of the measurement model. The extracted positions of image features are considered as camera measurements.

is obtained by the concatenation of the matrix $\boldsymbol{\Phi}^{ij} \in \mathbb{R}^{3\times 3}$ defined in (3.44) with zero matrix $\boldsymbol{0} \in \mathbb{R}^{3\times 3}$.

The physics-based approximation has a local character because only the model nodes in the neighborhood of the point under consideration are used for approximating the displacement of this point. It should be noted that the function $\underline{\chi}_k$ describing this approximation, is time-dependent. The reason for this is in the refinement of the spatial discretization of the model by inserting additional model nodes. This refinement will be introduced in Chapter 6. As a result, different model nodes can be located near the landmark $\underline{l}^i$ at different time steps.

Projection onto Image Plane of the Camera

Before presenting the second step of the derivation of the measurement model, the well-known homogeneous coordinates [80] are introduced for describing the central projection of the landmarks onto the image plane of the camera.

Transformation into Homogeneous Coordinates   The transformation of the Euclidean coordinates into homogeneous coordinates is done by adding the additional last coordinate. For example, mapping of the landmark position $\underline{l}_k^i$ into homogeneous coordinates is described by function

$$\underline{\varphi}^l : \underline{l}_k^i \mapsto w^i \left[\left(\underline{l}_k^i \right)^{\mathrm{T}} , 1 \right]^{\mathrm{T}} , \tag{3.59}$$

where w^i is arbitrary scalar. For the image feature $\underline{f}_k^{j,i}$, this mapping looks similarly

$$\underline{\varphi}^f : \underline{f}_k^{j,i} \mapsto v^{j,i} \left[\left(\underline{f}_k^{j,i} \right)^{\mathrm{T}} , 1 \right]^{\mathrm{T}} , \tag{3.60}$$

where $v^{j,i}$ is arbitrary scalar. For the back-transformation, the last coordinate of the vector in homogeneous coordinates should be removed. Accordingly, for the landmark $\underline{l}_k^i$ this transformation is defined by the function $\underline{\phi}^l : \mathbb{R}^4 \to \mathbb{R}^3$, which maps the four-dimensional vector in homogeneous coordinates in the three-dimensional one, whereby after eliminating the arbitrary scalar w^i, only the first three components of the homogeneous vector are considered. In the same way, when the position of the image feature $\underline{f}_k^{j,i}$ is of interest, the function $\underline{\phi}^f : \mathbb{R}^3 \to \mathbb{R}^2$ maps the three-dimensional vector in the two-dimensional one.

Mathematical Description of the Projection   The next step of the measurement model deduction is based on projecting the current position of the landmarks onto the images of every camera. This projection in its general form (3.56) is described by the function ζ. It should be noted that the measurement equation becomes nonlinear due to nonlinear mathematical formulation of the projection onto the image plane of the camera. The advantage of this is the omission of the computationally expensive three-dimensional reconstruction from camera images that occurs implicitly. Customarily, since the depth information is not explicitly reconstructed from the camera measurements, lower accuracy is expected along the camera axis. However, these inaccuracies are compensated by suitable non-parallel positioning of the cameras.

The relationship between the uncertain position of the landmark $\underline{l}_k^i$ and the uncertain position of image feature $\underline{f}_k^{j,i}$ is formulated by projecting the landmark onto the image plane of the camera

$$\underline{\varphi}^f\left(\underline{f}_k^{j,i}\right) = \mathbf{P}^{j,i}\underline{\varphi}^l\left(\underline{l}_k^i\right) , \tag{3.61}$$

where the projection matrix $\mathbf{P}^{j,i} = \mathbf{K}^j\left[\mathbf{R}^j\ \underline{t}^j\right]$ consists of the matrices of intrinsic $\mathbf{K}^j$ and extrinsic $\mathbf{R}^j$ camera parameters, provided by the calibration of the camera system, as well as of the translation vector $\underline{t}^j$. Then, as illustrated in Fig. 3.10, the measurement equation is obtained for every landmark

$$\underline{\hat{y}}_k^{j,i} := \underline{f}_k^{j,i} = \underline{\zeta}\left(\underline{\chi}_k\left(\underline{z}_k\right)\right) = \underline{\phi}^f\left(\mathbf{P}^{j,i}\underline{\varphi}^l\left(\underline{l}_k^i\right)\right) \tag{3.62}$$

by back-transformation of the image feature from the homogeneous coordinate system into the Cartesian one.

3.4.2 Measurement Uncertainties

Generally, there are two types of uncertainties in the measurement model: Systematic uncertainties, consisting of approximation and reprojection errors, and stochastic uncertainties incorporating the measurement noise that is caused, e.g., by uncertainties of the feature detection algorithm, electronic noise, changing light conditions, or flickering.

Systematic Uncertainties

The sources of the systematic uncertainties are twofold, as illustrated in Fig. 3.9. On one hand, they are introduced in the measurement model by a physics-based approximation, on the other hand, by projection of the landmarks positions onto the image plane of the camera.

Spatial Discretization Errors The physics-based approximation (3.57) exploits the numerical solution of the system of stochastic partial differential equations (3.24), which is corrupted by approximation errors introduced in Section 3.3.1. Therefore, the relationship (3.61) between the landmarks positions and image features is also afflicted with these errors. It should be noted that the a priori estimation of these errors is not necessary here because, similarly to the approximation errors of the system model, the approximation errors of the measurement model will be reduced in Chapter 6 based on the feedback from the visual motion compensation, which incorporates the a posteriori estimation of these errors.

Calibration Errors   Inaccuracies of calibration, such as incompletely eliminated distortion, uncertainties of the camera parameters or an inappropriate camera model, lead to reprojection errors $\underline{e}_k^{j,i} := [x_k^{e^{j,i}}, y_k^{e^{j,i}}]^{\mathrm{T}}$. Defined in pixel coordinates, these errors are represented by the distance between the true position of the image feature and the projection of the corresponding unknown true position of the landmark.

Stochastic Uncertainties

The stochastic uncertainties of the measurement model are caused by the measurement noise arising due to uncertainties of the feature detection algorithm, electronic noise, changing light conditions, or flickering. They are described by a random vector $\underline{v}_k^{j,i} = \left[x_k^{v^{j,i}}, y_k^{v^{j,i}} \right]^{\mathrm{T}}$, which probability density is assumed to be spatially and temporally white zero-mean Gaussian $\underline{v}_k^{j,i} \sim \mathcal{N}(\underline{0}, \boldsymbol{\Sigma}_k^{v^{j,i}})$ with covariance $\boldsymbol{\Sigma}_k^{v^{j,i}}$. The stochastic uncertainties are supposed to be independent and identically distributed.

Resulting Measurement Model

To summarize the results, the measurement model (3.1) is formulated for every camera j according to (3.57) and (3.62) in the form

$$
\begin{bmatrix} \hat{\underline{y}}_k^{j,1} \\ \hat{\underline{y}}_k^{j,2} \\ \vdots \\ \hat{\underline{y}}_k^{j,N} \end{bmatrix} = \begin{bmatrix} \phi^f \left(\mathbf{P}^{j,1} \underline{\varphi}^l \left(\underline{l}^1 + \mathbf{H}^1 \underline{z}_k \right) \right) \\ \phi^f \left(\mathbf{P}^{j,2} \underline{\varphi}^l \left(\underline{l}^2 + \mathbf{H}^2 \underline{z}_k \right) \right) \\ \vdots \\ \underline{\phi}^f \left(\mathbf{P}^{j,N} \underline{\varphi}^l \left(\underline{l}^N + \mathbf{H}^N \underline{z}_k \right) \right) \end{bmatrix} + \begin{bmatrix} \underline{e}_k^{j,1} \\ \underline{e}_k^{j,2} \\ \vdots \\ \underline{e}_k^{j,N} \end{bmatrix} + \begin{bmatrix} \underline{v}_k^{j,1} \\ \underline{v}_k^{j,2} \\ \vdots \\ \underline{v}_k^{j,N} \end{bmatrix} , \quad (3.63)
$$

where the vector $\underline{z}_k \in \mathbb{R}^{6N_M}$ denotes the system state. The size of the system state is determined by the number of the model nodes N_M and their six degrees of freedom. N is specified by the amount of image features $\hat{\underline{y}}_k^{j,i} \in \mathcal{Y}_k^j$ that are extracted from the image acquired by camera j at time step k and have unique correspondences in the images of the other two cameras. The elements of the matrix

$$
\mathbf{H}^i = \left[\mathbf{H}^{i1}, \mathbf{H}^{i2}, \ldots, \mathbf{H}^{iN_M} \right] \tag{3.64}
$$

are matrices $\mathbf{H}^{ij} \in \mathbb{R}^{3\times6}$ defined in (3.58). The reprojection errors $\underline{e}_k^{j,i}$ ordered in the vector $\underline{e}_k^{j}$ represent the systematic errors of the measurement model. The stochastic errors $\underline{v}_k^{j,i}$ collected in the vector $\underline{v}_k^{j}$ are characterized by the measurement noise.

3.5 Summary

The aim of this chapter, which is fundamental for estimating the heart motion, is twofold. First of all, it proposes the methodology for constructing an appropriate physical model of the beating heart in order to understand its complex behavior. Then, it deals with the derivation of the discrete state-space model, consisting of the system and measurement models, from this model.

Contributions The novelty of the proposed physical model is in the volumetric representation of the heart wall. Although volumetric models of the heart are widespread in other applications, like preoperative planing or diagnostics, to the best of our knowledge, no such models have been proposed for beating heart surgery. The major braking factor here is the high computational burden of these models. However, compared to surface models, they allow to reproduce the volumetric behavior of the heart. Basically this means that the motion of every point of the heart is influenced by surrounding points on the surface of the heart and in its interior.

The main challenge here is to balance the computational complexity of the model and its accuracy. To face this problem, the physical model considers only the heart wall within the area of the beating heart, where surgical interventions are performed. This gives rise to mathematically challenging definitions of boundary conditions, which have to represent accurately the connection of this part of the heart with remaining heart tissues. The motion of the heart wall is described by a distributed-parameter system that is derived based on physical principles underlying the motion of the heart. This yields the system of hyperbolic stochastic partial differential equations. Due to the combination of the Hooke's law with linear strain tensor for the description of the heart wall deformation, this system is linear.

With respect to the state-space model, the contribution of this chapter concerns the exploitation of the known physical properties of the object. Another important aspect of this model is the combination of a simplified description of the underlying physical object with the consideration of stochastic and systematic errors of model at each stage of the state-space model deduction.

The general procedure for deriving the physics-based system model is similar and typical for the numerical solution of partial differential equations. The application of the method of lines allows for separate discretization of this system in space and time. For spatial discretization, the meshless local Petrov-Galerkin mixed collocation method is chosen. This method refers to the group of the element-free methods, which are still not so common in modeling of deformable objects. Compared to classical element-based methods, this method discretizes the spatial domain of the model by a set of nodes without defining their connectivity. Such an element-free discretization yields better accuracy of solution [126], requires less computational resources and proposes higher flexibility in model refinement. As a result of the spatial discretization, the random fields of the system are represented by a finite number of time-continuous random variables. Subsequently, the implicit Euler method, which is unconditionally stable, is applied for temporal discretization.

The measurement model, involving the numerical solution of the system of stochastic partial differential equations, is deduced in two steps. At first, the current position of the heart surface is obtained by a physics-based approximation, so named thanks to its parameters determined in a physically correct fashion. Then, this position is projected onto the image plane of the camera, yielding the corresponding image feature in the camera image.

Further Developments   Certain open issues still remain. First of all, the proposed physical model is object-specific. This gives rise to the problem of identifying the physical properties of the heart wall in order to perform an accurate estimation of its motion. This problem can be addressed by various methods that attempt to directly measure tissue parameters [35, 135, 219], or determine them by fitting the model to available measurement data [106, 166, 188]. The drawback of these methods is the neglect of measurement uncertainties. This is the reason that in this thesis, the model parameter will be estimated by a nonlinear filtering approach introduced in Chapter 4. For efficiency of estimation, this estimator splits the system state into linear and nonlinear substructures. However, in this way, the problem of high dimensionality of the state will not be alleviated. For solution of this problem, model order reduction techniques, such as high-order Guyan dynamic condensation or modal-type condensation proposed in [167], can serve as a starting point. Furthermore, it should be noted that a prerequisite for the measurement model is that the correspondences between the image features are known. In practice, the extraction of this information is a challenging task that will be tackled in the next chapter.

With regard to further development of the models, it should be noted that the proposed physical model builds a foundation for modeling of surgical interventions. During beating heart surgical procedures, the heart undergoes extensive surgical manipulations, such as heart rotation for circumflex artery anastomosis, or delicate maneuvers, such as sewing and cutting. However, to the best of our knowledge, no models handling the surgical interventions, have been proposed for beating heart surgery. Generally, the consideration of these manipulations will lead to treating the model discontinuities wherefby such methods as visibility method, diffraction method, or transparency method [152] can be applied. In contrast to existing surface models, the proposed volumetric model permits the modeling of penetrations in interior of the organ, because it is not empty inside.

Another point of the future work is the extension of the model with respect to a more sophisticated representation of the myocardium's structure. This may become essential for realistic modeling of surgical interventions. One of the possible representations of the laminar and fiber-filled organization of cardiac muscle is a composite material [47, 55], consisting of material layers with different physical properties. One of the main challenges here is an accurate identification of the physical properties of individual layers, e.g., by using medical imaging modalities, like computer tomography or ultrasound. For dropping severe restrictions on small deformations that can lead to more realistic representation of the heart tissues, nonlinear strain tensors, like Green and St. Venant strain tensor [53], can be used. Here, special care must be taken of the fact that the model will be described by a system of nonlinear partial differential equations. This will lead to substantial complication of numerical calculations and therefore, increasing computational complexity.

Employing models of heart excitation promises a more rapid and perspective reaction of the models on the changes of the heart motion. First of all, motivated by known pressure and volume of the left ventricle, which can be accessed by a heart catheter, mathematical models of pressure-volume relation, e.g., time varying elastance model [203], can provide the input information for the physical model of the heart wall. A more elegant solution is to take into account an electrical activity of the heart. The common models of this activity are FitzHugh-Nagumo equations [163] and models based on myofibre active constitutive law [30, 179, 188]. As electrocardiogram signal, measuring electrical activity and muscular functions of the heart, allows to detect an abnormal motion of the heart about 90 ms ahead, as stated in [57], its incorporation in the physical model would significantly improve the model's abilities. With respect to the motion compensation in a robotic system for beating heart surgery, it should be noted that this signal has been used in [57, 159] for robust motion prediction, also in combination with other measurement modalities.

4 Physics-Based Tracking of Deformable Object

This chapter is concerned with the visual tracking of an elastically deformable object as observed by a trinocular camera system. Formulated as a probabilistic problem, this tracking provides statistical information about the most probable three-dimensional positions of some points on the surface of the object.

With regard to the beating heart robotic surgery system introduced in Section 1.1, the visual tracking of a deformable object, such as a beating heart, is essential for the main tasks of the system: synchronization, navigation, manipulation, and visual motion compensation. Accordingly, for synchronizing the surgical instruments with the beating heart, the predicted positions of the heart landmarks are adopted for control of the surgical robot. Furthermore, for planning and fulfillment of robotic manipulations, such as the cutting of heart tissues, or sewing, the most accurate estimation of the current position of the heart tissues is important for the safety of the surgical operations. In addition, for visual motion compensation, the heart position is necessary for compensating the changes arising in camera images due to deformation of the heart.

Commonly, the methods for visual tracking reconstruct only the surface of the deformable object [173, 174]. The motion inside the object remains inaccessible. However, this valuable information would sufficiently extend the capabilities of a surgeon in an intraoperative planing of robotic manipulations and navigation. While existing works focus on a registration of preoperative models with intraoperative data [87, 104, 166], in attempting to perform the registration very rapidly, the methods proposed in this chapter, enable a runtime estimation of the entire heart wall deformation from camera images. This will be demonstrated in evaluation results presented in Chapter 7.

There are three further points that differentiate the proposed physics-based tracking from most of the existing methods for tracking of the heart motion that are introduced in Section 1.3. First of all, the tracking operates directly in a physical space. Based on the physics-based state-space model presented in Chapter 3, it predicts and estimates the heart position in a physically correct way. The second point is that the proposed tracking is probabilistic. Targeted at extracting the related information from noisy measurement data, it considers systematic and stochastic errors of the model and measurements by means of nonlinear stochastic estimation. The third point lies in exploiting the physically correct prediction of the heart position for extraction of the measurements from camera images. In particular, beyond multiple gating criteria for establishing the correspondences between the image features and landmarks, the physics-based criterion is introduced, which filters out the measurement outliers.

Before illustrating the key idea in Section 4.2, firstly, in Section 4.1, the tracking is formulated as a probabilistic problem of estimating the most probable positions of the heart landmarks at every time step. Then, Section 4.3 deals with adopting the physically justified prediction of the landmark positions for extraction of measurement information from the camera images. In Section 4.4, after an initialization of the models, these positions are reconstructed by the physics-based approximation, which is combined with the nonlinear stochastic estimation of the system state, parameters, and systematic errors. Finally, this chapter concludes with the discussion of the main points of the proposed tracking method.

The tracking approach introduced in this chapter was primarily published in [240] and further extended to the estimation of the model parameters in [239]. This chapter expands the results of the papers to the handling of the modeling and measurement errors that are systematic.

4.1　Problem Formulation

The visual tracking deals with detecting the three-dimensional positions of the heart landmarks by extracting the related information from camera images acquired at consecutive time steps. In contrast to the variety of the deterministic methods for tracking the heart surface motion [57, 160, 182] that do not consider the uncertainties at all, in this section, the tracking is formulated as a probabilistic problem. The main advantage of such formulation lies in rejecting the measurement disturbances by utilizing a priori information about the behavior of the object. The two main issues of this problem consist in gaining measurement information and estimation of landmark positions.

Extraction of Measurement Information   One of the main problems of the heart motion tracking is that the camera measurements are not directly available. Therefore, they should be extracted from the camera images provided by a camera system consisting of n cameras. It must be highlighted that the set of measurements $\mathcal{Y}_k = \bigcup_{j=1}^{n} \mathcal{Y}_k^j$ incorporates only image features $\mathcal{Y}_k \subseteq \mathcal{F}_k$ from set $\mathcal{F}_k = \bigcup_{j=1}^{n} \mathcal{F}_k^j$ that can be identically related to the landmarks under observation at the same time step. Here, the order of elements in the sets $\mathcal{L}_k$ and $\mathcal{F}_k$ is fixed. Although due to overlapping coverage of the n cameras the same landmarks $\mathcal{L}_k$ are observed, every set $\mathcal{F}_k^j$ originated from camera j may include a different amount of features. The reason is that features can be lost in every camera image, e.g., due to occlusions, and also false detections, e.g., due to illumination artifacts, are possible. Therefore, the one of the important aims of the tracking is to find the correspondences between the image features $\underline{f}_k^{j,i} \in \mathcal{F}_k^j$ and landmarks $\underline{l}_k^i \in \mathcal{L}_k$ at every time step.

Estimation of Landmark Positions   While the camera measurements are noisy, the tracking of the landmark positions can be formulated as a probabilistic problem related to a Bayesian estimation. According to Bayes' theorem, e.g., introduced in [125], the current positions of the landmarks are characterized by a conditional density

$$f\left(\underline{l}_k^i \mid \mathcal{Y}_{1:k}\right) = \alpha_k f^L\left(\mathcal{Y}_k \mid \underline{l}_k^i\right) f\left(\underline{l}_k^i \mid \mathcal{Y}_{1:k-1}\right),$$

which depends on likelihood $f^L\left(\mathcal{Y}_k \mid \underline{l}_k^i\right)$ processing the current observations $\mathcal{Y}_k$ and a normalization factor α_k. It should be noted that one of the conditional densities

$$f_k^p\left(\underline{l}_k^i\right) := f\left(\underline{l}_k^i \mid \mathcal{Y}_{1:k-1}\right), \; f_k^e\left(\underline{l}_k^i\right) := f\left(\underline{l}_k^i \mid \mathcal{Y}_{1:k}\right)$$

denotes the a priori $f_k^p\left(\underline{l}_k^i\right)$ density of the landmark position characterized by an estimate from the previous time steps. The other conditional density $f_k^e\left(\underline{l}_k^i\right)$ represents the a posteriori density characterizing an estimate updated by a measurement information at the current time step. Evidently, both densities depend on the union of sequences of camera system measurements. Accordingly, the a posteriori density depends on the sequence

$\mathcal{Y}_{1:k} = \bigcup_{j=1}^{n} \mathcal{Y}_{1:k}^{j}$, which assembles the measurements $\mathcal{Y}_{1:k}^{j} = \left\{ \mathcal{Y}_1^{j}, \ldots, \mathcal{Y}_k^{j} \right\}$ of each camera j acquired up to respective time step.

If the Markov property of the system is to be presumed, the densities of the estimates

$$f_k^e \left(\underline{l}_k^i \right) = f \left(\underline{l}_k^i \mid \mathcal{Y}_k \right) , \quad f_k^p \left(\underline{l}_k^i \right) = f \left(\underline{l}_k^i \mid \mathcal{Y}_{k-1} \right)$$

depend only on the last available observations. Consequently, the a posteriori density of the landmark position can be obtained at every time step by the Chapman-Kolmogorov equation, e.g., introduced in [110]

$$\begin{aligned} f_k^e \left(\underline{l}_k^i \right) &= \alpha_k f^L \left(\mathcal{Y}_k \mid \underline{l}_k^i \right) f_k^p \left(\underline{l}_k^i \right) \\ &= \alpha_k f^L \left(\mathcal{Y}_k \mid \underline{l}_k^i \right) \int_{\mathbb{R}^3} f^T \left(\underline{l}_k^i \mid \underline{l}_{k-1}^i \right) f_{k-1}^e \left(\underline{l}_{k-1}^i \right) \mathrm{d}\underline{l}_{k-1}^i , \end{aligned} \quad (4.1)$$

where $f^T \left(\underline{l}_k^i \mid \underline{l}_{k-1}^i \right)$ is the transition density derived from the system model.

There are some challenges in the computation of the a posteriori density. First of all, the Chapman-Kolmogorov equation (4.1) is not analytically solvable because of the nonlinearity of the state-space model derived in Chapter 3. Furthermore, the system and measurement models, which are used for derivation of the transition density $f^T \left(\underline{l}_k^i \mid \underline{l}_{k-1}^i \right)$ and likelihood $f^L \left(\mathcal{Y}_k \mid \underline{l}_k^i \right)$, are not exactly known. They incorporate unknown physical parameters and modeling errors. Moreover, the problem intensifies if the landmark positions have to be accurately determined when none or only some of the measurements are available, e.g., in case of occlusions.

4.2   Tracking Overview

The proposed tracking schematically illustrated in Fig. 4.1 incorporates two key ideas. The first idea consists in physically reasonable approximation of the densities $f_k^e \left(\underline{l}_k^i \right)$ and $f_k^e \left(\underline{l}_k^i \right)$, which describe the a priori and a posteriori estimates of the landmark position. The second idea behind the tracking is to exploit the a priori known physical information for gaining measurement information from camera images.

Preliminaries   For the explanation of the physically reasonable approximation of the landmark positions, it should be recalled that the elements

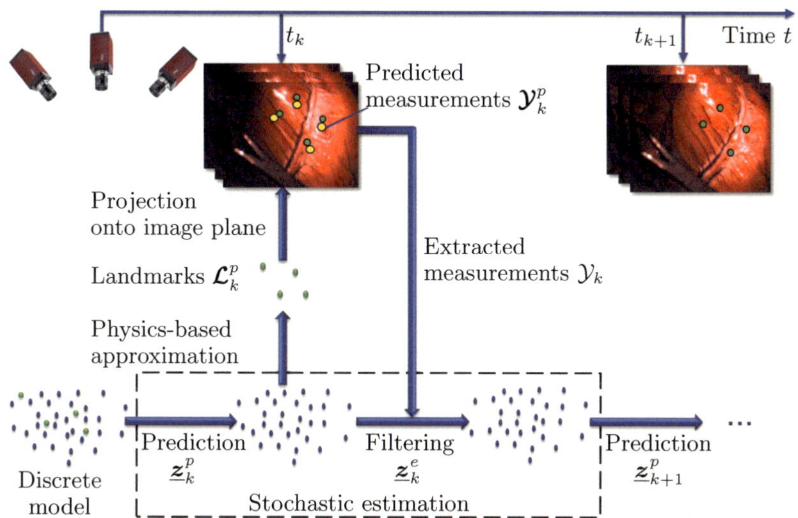

Figure 4.1: Key ideas of the tracking approach. The system state is propagated over time by a nonlinear stochastic estimation based upon physics-based state-space model. This model incorporates the physical knowledge about object under observation in the tracking. For extracting the measurement data $\mathcal{Y}_k$ from camera images the physically reasonable a priori state estimate $\underline{z}_k^p$ is used.

of the system state $\underline{z}_k^i \in \mathbb{R}^6$ collected in the vector $\underline{z}_k \in \mathbb{R}^{n_{z_k}}$ are assigned to the set of model nodes in the undeformed configuration of the physical model $\mathcal{M} = \left\{\underline{m}^i\right\}_{i=1}^{N_M}$. Therefore, the size of the state vector $n_z = 6N_M$ is determined by the number of model nodes N_M and 6 degrees of freedom of the vector $\underline{z}_k^i$.

It is worth mentioning that the relationship between the system state and the landmarks is established by a physics-based approximation defined by the time-dependent function $\underline{\chi}_k$ in (3.55) and (3.57).

Furthermore, the projection of the landmark positions $\underline{l}_k^i \in \mathcal{L}$ onto an image plane of a camera is described by the function $\underline{\zeta}$ introduced in (3.56) and (3.62).

Physically Reasonable Estimates According to physics-based approximation (3.57), the a priori $f_k^p\left(\underline{l}_k^i\right)$ and a posteriori $f_k^e\left(\underline{l}_k^i\right)$ densities of the landmark position are determined by

$$
f_k^p\left(\underline{l}_k^i\right) = \int\limits_{\mathbb{R}^{n_{z_k}}} f^R\left(\underline{l}_k^i \mid \underline{z}_k\right) f_k^p\left(\underline{z}_k\right) \mathrm{d}\underline{z}_k \,, \quad f_k^e\left(\underline{l}_k^i\right) = \int\limits_{\mathbb{R}^{n_{z_k}}} f^R\left(\underline{l}_k^i \mid \underline{z}_k\right) f_k^e\left(\underline{z}_k\right) \mathrm{d}\underline{z}_k \,,
$$

$$(4.2)$$

where the conditional density $f^R\left(\underline{l}_k^i \mid \underline{z}_k\right)$ is the transition density of the relationship between the system state and the landmark. It is deduced from the physics-based approximation function $\underline{\chi}_k$. The a priori $f_k^p\left(\underline{z}_k\right)$ and a posteriori $f_k^e\left(\underline{z}_k\right)$ densities of the system state are determined by using the system state in Chapman-Kolmogorov equation (4.1) that leads to

$$
\begin{aligned}
f_k^e\left(\underline{z}_k\right) &= \alpha_k f^L\left(\mathcal{Y}_k \mid \underline{z}_k\right) f_k^p\left(\underline{z}_k\right) \\
&= \alpha_k f^L\left(\mathcal{Y}_k \mid \underline{z}_k\right) \int\limits_{\mathbb{R}^{n_{z_k}}} f^T\left(\underline{z}_k \mid \underline{z}_{k-1}\right) f_{k-1}^e\left(\underline{z}_{k-1}\right) \mathrm{d}\underline{z}_{k-1}.
\end{aligned}
$$

$$(4.3)$$

It should be noted that from (4.2) and (4.3) follows the dependence of the a priori $\underline{l}_k^{i,p} \sim f_k^p\left(\underline{l}_k^i\right)$ and a posteriori $\underline{l}_k^{i,e} \sim f_k^e\left(\underline{l}_k^i\right)$ positions of every landmark $\underline{l}_k^i \in \mathcal{L}_k$ from the physics-based state-space model introduced in Chapter 3. In particular, the transition density $f^T\left(\underline{z}_k \mid \underline{z}_{k-1}\right)$ is derived from the system model presented in Section 3.3, which is parameterized by the physical properties of the object under observation, such as, e.g., the Young modulus E and the Poisson ratio ν. Furthermore, the likelihood $f^L\left(\mathcal{Y}_k \mid \underline{z}_k\right)$ is deduced from the measurement model proposed in Section 3.4. Therefore, it also incorporates the physical information inherited from the physics-based approximation of the current position of the object.

Due to nonlinearity of the state-space model, the densities of the system state are not analytically determinable. For its numerical approximation, diverse nonlinear stochastic estimators, e.g., overviewed in [194, 195, 223] can be applied. They approximates the densities of the system state in two steps: prediction step and filter step. The prediction step provides the a priori state $\underline{z}_k^p \sim f_k^p\left(\underline{z}_k\right)$ based on the past data. In the filter step, the a posteriori state $\underline{z}_k^e \sim f_k^e\left(\underline{z}_k\right)$ is obtained by updating the a priori state by current camera measurements $\mathcal{Y}_k$.

Physics-Based Gating   The second idea of the tracking is to link the image features extracted from camera images with the heart landmarks by exploiting physically reasonable a priori state estimate $\underline{z}_k^p$. For constructing the measurement vector $\mathcal{Y}_k$, it must be determined which image features result from which landmarks.

In order to obtain this information, multiple gating criteria are applied. One of the gating functions is determined by Mahalanobis distance [136], which defines the regions where image features associated with the certain landmarks can be found with high probability. For this purpose, the a priori positions of landmarks $\underline{l}_k^{i,p} \in \mathcal{L}_k^p$, the densities (4.2) of which depend on physics-based approximation function $\underline{\chi}_k$, are projected onto the image planes.

Interestingly, the gating implies physical information although the applied multiple gating criteria are typical for visual tracking. This is thanks to physical model incorporated in the tracking. The a priori positions of the landmarks $\underline{l}_k^{i,p}$ are a priori estimates of the solution of the system of stochastic partial differential equations (3.22) that describes this physical model.

4.3 Physics-Based Extraction of Measurement Information

This section is concerned with the extraction of the measurement information from camera images by exploiting the a priori known physical information. The focus is on establishing the correspondences between the image features and landmarks in order to enable the processing of the measurements by an estimator. After formulating the correspondence problem for physics-based tracking, a correspondence function for linking the image features with the landmarks is constructed. Subsequently, the four multiple gating criteria incorporated in this function are introduced.

4.3.1 Formulation of Correspondence Problem

The central task of the correspondence problem is to figure out the correct mapping between the landmarks $\mathcal{L}_k$ and measurements $\mathcal{Y}_k$ that are extracted from images acquired by n different cameras at the same time step. These measurements are selected from the sets of the image features $\mathcal{F}_k = \bigcup_{j=1}^{n} \mathcal{F}_k^j$, each of these $\mathcal{F}_k^j$ can involve measurement outliers and artifacts. Moreover, some of the features can be lost or may not be detected.

Therefore, the correspondence problem consists in assignment of each image feature from the set $\mathcal{F}_k^j$ to at most one of the landmarks from the set $\mathcal{L}_k$. Furthermore, it must be assured that each landmark gets no more than one measurement in every set $\mathcal{F}_k^j$.

The a posteriori density of the landmark position is obtained by plugging (4.2) in (4.3) that yields

$$f_k^e\left(\underline{l}_k^i\right) = \int\limits_{\mathbb{R}^{n_z}} \alpha_k f^L\left(\mathcal{Y}_k, \underline{l}_k^i \mid \underline{z}_k\right) f_k^p\left(\underline{z}_k\right) \mathrm{d}\underline{z}_k . \tag{4.4}$$

In this equation, the joint likelihood function $f^L\left(\mathcal{Y}_k, \underline{l}_k^i \mid \underline{z}_k\right)$ determines the hypothetical probability that the system state $\underline{z}_k$ would yield the position of the landmark $\underline{l}_k^i \in \mathcal{L}_k$ and measurements $\mathcal{Y}_k$. In doing so, the measurements $\mathcal{Y}_k$ and landmark position $\underline{l}_k^i$ are assumed to be conditionally independent. It should be noted that in order to determine the densities of the system state $\underline{z}_k$ by means of stochastic estimation, the correspondences between the acquired measurements $\mathcal{Y}_k$ and landmarks $\mathcal{L}_k$ must be known.

Since the state and, therefore, the landmark positions, as well as positions of the image features, are noisy, there can be a large amount H of possible hypotheses of which image features correspond to which landmarks. In order to consider this, according to [20, 137], the joint likelihood function should be marginalized across all possible hypotheses. Therefore, from (4.4) follows

$$f^L\left(\mathcal{Y}_k, \underline{l}_k^i \mid \underline{z}_k\right) = \kappa \sum_{h=1}^{H} f\left(\mathcal{F}_k, \underline{l}_k^i, \mid \underline{z}_k, \mathbf{D}_k^h\right) \approx \kappa f\left(\mathcal{F}_k, \underline{l}_k^i, \mid \underline{z}_k, \mathbf{D}_k\right) , \tag{4.5}$$

where κ is the normalization constant and matrix $\mathbf{D}_k^h$ defines which measurements originate from which landmarks.

As the handling of a large amount of hypotheses will strongly affect the real-time operability of the system, especially while the physics-based state-space model is high-dimensional and nonlinear, such tracking can hardly be applied in a beating heart surgery system. Therefore, as proposed in [20, 199], all hypotheses are approximated in (4.5) by one best hypothesis, described by the matrix $\mathbf{D}_k$. The elements of the matrix $\mathbf{D}_k = \left[\underline{d}^1, \dots, \underline{d}^{N_L}\right]$ are assignment vectors $\underline{d}_k^i$ containing n indices of

---

Algorithm 1 Physics-based gating of image features $\mathcal{F}_k^1, \ldots, \mathcal{F}_k^n$ at time step t_k

---

Require: A priori position of the landmark $\underline{l}_k^{i,p} \in \mathcal{L}_k$
  1: Evaluate I_E // Epipolar criterion
  2: Evaluate I_P for $\underline{l}_k^{i,p} \in \mathcal{L}_k$ // Physics-based criterion
  3: **if** $I_C \mapsto 0$ **then**
  4:     Evaluate I_T // Triangulation criterion
  5:     Evaluate I_C // Consistency criterion
  6: **end if**
  7: Compute assignment vector $\underline{d}_k^i$

---

the image features. These features, uniquely corresponding to the ith of N_L landmarks, are selected from sets $\mathcal{F}_k^j$ of every jth of n cameras.

4.3.2 Physics-Based Correspondence Function

This section aims at establishing the best hypothesis that determines which measurements originate from which landmarks. This hypothesis will be determined by the correspondence function that incorporates four gating criteria. Beyond the well-known epipolar, triangulation, and consistency criteria, a physics-based criterion using Mahalanobis distance is introduced.

The assignment vectors assembling the matrix $\mathbf{D}_k$ are defined by a maximum a priori estimate

$$\underline{d}_k^i = \arg\max_{i \in \mathbb{Z}_+} \left(f\left(\mathcal{F}_k^1, \ldots, \mathcal{F}_k^n \mid \underline{l}_k^i \right) \right) \tag{4.6}$$

that maximizes probability that some of the image features $\mathcal{F}_k^j$ extracted from the jth image of the camera are true correspondences of the landmark $\underline{l}_k^i \in \mathcal{L}_k$. For gating the image features, the probability $f\left(\mathcal{F}_k^1, \ldots, \mathcal{F}_k^n \mid \mathcal{L}_k \right)$ is determined by correspondence function that incorporates a series of conditions

$$f\left(\mathcal{F}_k^1, \ldots, \mathcal{F}_k^n \mid \underline{l}_k^i \right) = I_E\left(\gamma_E \right) \cdot I_P\left(\gamma_P \right) \cdot I_C \cdot I_T \cdot I_C . \tag{4.7}$$

The computation of this function is listed in Algorithm 1. The first condition for gating the image features is represented by an epipolar criterion

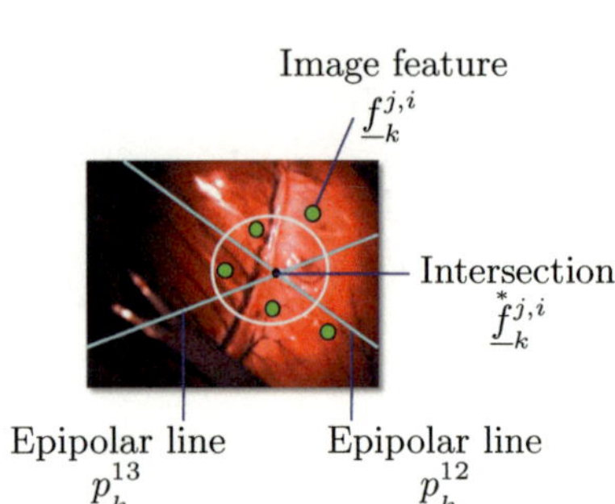

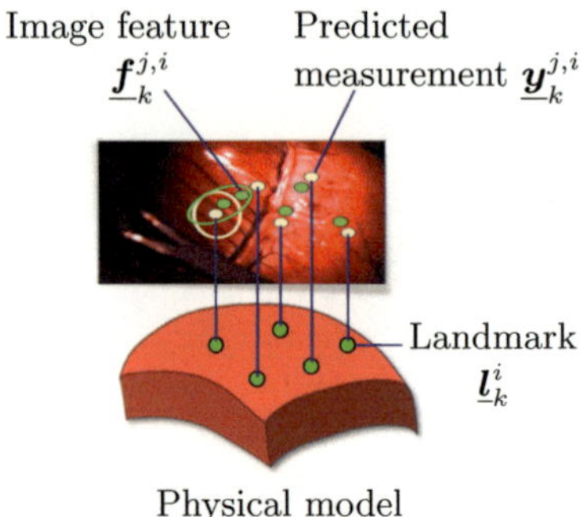

(a) Epipolar criterion. The projection of the certain landmark in one of the camera images is determined by the intersection of epipolar lines originated from projections of the same landmark in other two views. Due to calibration inaccuracies, the image features within a limited radius from the cross of epipolar lines are considered as possible correspondences.

(b) Physics-based criterion considers only those image features as being corresponding to the landmarks which are near their expected physically reasonable location determined by the predicted measurements. As the predicted measurements and image features are uncertain, their variances are illustrated by ellipsoids.

Figure 4.2: Epipolar and physics-based criteria for reducing the number of the image features, which may be originated from a certain landmark on the heart surface.

(line 1) defined by an indicator function I_E depending upon the gating parameter γ_E. Furthermore, the physics-based criterion (line 2), defined by a function I_P depending on gating parameter γ_P is introduced. It should be noted that when the consistency criterion I_C (line 3) evaluated before the triangulation criterion I_T (line 4) is satisfied, the calculation of the triangulation criterion, is not necessary. The consistency criterion ensures the uniqueness of correspondences.

Epipolar Criterion

The epipolar criterion aims at reducing the number of image features $\underline{f}_k^{j,i} \in \mathcal{F}_k^j$ extracted from images of every jth camera, which may be originated from a certain landmark $\underline{l}_k^i \in \mathcal{L}_k$. This criterion is based on the fact [80] that the projection of the landmark $\underline{l}_k^i$ onto the image plane of one of the cameras $\underline{f}_k^{j,i}$, $j = 1, \ldots, n$ is determined by the intersection of epipolar lines, originated from projections of the same landmark onto the image planes of other two cameras. This means, for example, that the image feature $\underline{f}_k^{1,i} \in \mathcal{F}_k^1$ in the image of the first camera should be placed at position

$$\overset{*}{\underline{f}}_k^{1,i} := \phi^f \left(\underline{p}^{12} \times \underline{p}^{13} \right) = \phi^f \left(\left(\mathbf{F}_{12}\, \varphi^f \left(\underline{f}_k^{2,i} \right) \right) \times \left(\mathbf{F}_{13}\, \varphi^f \left(\underline{f}_k^{3,i} \right) \right) \right) \quad (4.8)$$

denoted by the intersection of epipolar lines $\underline{p}^{12}$ and $\underline{p}^{13}$, as illustrated in Fig. 4.2(a). These lines are obtained by projecting the epipolar lines, passing through the image features $\underline{f}_k^{2,i} \in \mathcal{F}_k^2$ and $\underline{f}_k^{3,i} \in \mathcal{F}_k^3$ on the images of the second and third cameras, onto the image plane of the first camera. In the equation (4.8), the function φ^f converts the position of the image feature into homogeneous coordinates, as introduced in (3.60), whereby the function ϕ^f makes the inverse mapping. The fundamental matrices $\mathbf{F}_{1j}$ are determined according to [80] by the equation

$$\mathbf{F}_{1j} = [e]_\times \, \mathbf{P}^j \mathbf{P}^+ .$$

Here, the matrix $\mathbf{P}^j$ is the projection matrix of the jth camera, $\mathbf{P}^+$ is a pseudo-inverse of the projection matrix of the first camera, and the skew symmetric matrix

$$\left[e^j \right]_\times = \begin{bmatrix} 0 & -z^{e^j} & y^{e^j} \\ z^{e^j} & 0 & -x^{e^j} \\ -y^{e^j} & -x^{e^j} & 0 \end{bmatrix}$$

is formed from the components of the epipole vector $\underline{e}^j = \left[x^{e^j}, y^{e^j}, z^{e^j} \right]^{\mathrm{T}}$. This vector is defined by the intersection of the baseline with the image plane of the camera.

However, as the camera measurements are corrupted by uncertainties and the projection matrices are inaccurate due to calibration errors, the image features extracted from camera images do not really coincide with their approximated position. Therefore, the image features within a very limited radius γ_E from the intersection of epipolar lines passing through two image features in the other two camera views, are considered as possible matches of these features. This is defined by the indicator function

$$I_E\left(\gamma_E\right) := \begin{cases} 1 & \text{if } \sum\limits_{j=1}^{n} \left\| \overset{*}{\underline{f}}_k^{j,i} - \underline{f}_k^{j,i} \right\| \leq \gamma_E, \\ 0 & \text{otherwise,} \end{cases}$$

where the notation $\| \cdot \|$ denotes the Euclidean norm on $\mathbb{R}^2$.

It should be noted that the well-known deficiency of this criterion is that the two epipolar lines on the first image may coincide by improper configurations of the camera system. In this case, the position of the image feature cannot be approximated by (4.8). In order to avoid this, criteria using a trifocal tensor can be applied [80]. However, as in this thesis the cameras are placed in such way that maximum accuracy can be achieved and no degenerated configurations occur, the position of the image features is completely determined by (4.8). In addition, the evaluation of this criterion has lower computational complexity than using criteria based on trifocal tensor.

It can happen that more than one image features are near the intersection of the epipolar lines, as shown in Fig. 4.2(a). This is especially often the case when the camera calibration has low accuracy, so that fundamental matrices are significantly affected by errors. In order to resolve this problem, other criteria will be applied in the following section.

Physics-Based Criterion

In the further step, from the image features satisfying the epipolar criteria only those are selected that pass through physics-based criterion. Based on generic gating procedure for multi-target tracking [20, 125], this criterion considers only those image features as being corresponding to the landmarks that are near their expected physically reasonable location.

The key idea behind this criterion is depicted in Fig. 4.2. First of all, the stochastic filter that will be presented in Section 4.4.3, predicts the positions of the image features $\underline{f}_k^{j,i} \in \mathcal{F}_k^j$ corresponding to the landmarks $\underline{l}_k^i \in \mathcal{L}_k$. For this purpose, it evaluates the measurement equation (3.63). Hence, the a priori positions of the landmarks $\underline{l}_k^{i,p} \in \mathcal{L}_k^p$ approximated in a physically correct way are projected onto the image plane of a camera, as schematically illustrated in Fig. 4.1. This yields the expected positions of the image features, also called predicted measurements $\underline{y}_k^{j,i} \in \mathcal{Y}_k^p$, those correspondences to the landmarks are known. Here, these positions as well as the positions of the image features $\underline{f}_k^{j,i} \in \mathcal{F}_k^j$ extracted from the camera image, denoted by

$$\underline{y}_k^{j,i} \sim f_k\left(\underline{y}_k^{j,i}\right) \approx \mathcal{N}\left(\underline{\mu}_k^{y^{j,i}}, \mathbf{\Sigma}_k^{y^{j,i}}\right), \quad \underline{f}_k^{j,i} \sim f_k\left(\underline{f}_k^{j,i}\right) \approx \mathcal{N}\left(\underline{\mu}_k^{f^{j,i}}, \mathbf{\Sigma}_k^{f^{j,i}}\right),$$

are assumed to be statistically independent and Gaussian with respective mean and covariance functions.

Subsequently, a region $\mathcal{R}$ is defined in the camera image, where the image features corresponding to the certain landmark are located with high probability

$$\mathcal{R}\left(\gamma_P\right) := \left\{ \underline{f}_k^{j,i} : \underline{d}_M^{j,i} \leq \gamma_P \right\}.$$

This region is determined by a squared Mahalanobis distance [136] according to

$$\underline{d}_M^{j,i} := \left(\underline{\mu}_k^{f^{j,i}} - \underline{\mu}_k^{y^{j,i}}\right)^{\mathrm{T}} \left(\boldsymbol{\Sigma}_k^{j,i}\right)^{-1} \left(\underline{\mu}_k^{f^{j,i}} - \underline{\mu}_k^{y^{j,i}}\right),$$

where the inverse of the positive-definite matrix is given by

$$\left(\boldsymbol{\Sigma}_k^{j,i}\right)^{-1} = \left(\boldsymbol{\Sigma}_k^{y^{j,i}} + \boldsymbol{\Sigma}_k^{f^{j,i}}\right)^{-1}.$$

The gating parameter γ_P is chi-square distributed. It is determined by a table of the chi-square distribution [20] with the number of degrees of freedom $d = 2$, which is equal to the dimension of the image feature. This table gives the probability

$$f\left(\underline{f}_k^{j,i} \mid \underline{l}_k^i\right) = f\left(\underline{f}_k^{j,i} \in \mathcal{R}\left(\gamma_P\right)\right)$$

that the image feature $\underline{f}_k^{j,i}$ is in the region $\mathcal{R}$, which is equal to the probability that this image feature corresponds to the landmark $\underline{l}_k^i$. Therefore, for selecting only the image features that most probably correspond to the landmark the gating function is defined by this probability

$$I_P\left(\gamma_P\right) := \begin{cases} f\left(\underline{f}_k^{j,i} \mid \underline{l}_k^i\right) & \text{if } \underline{d}_M^{j,i} \leq \gamma_P \\ 0 & \text{otherwise}. \end{cases} \tag{4.9}$$

In this way, the image features located in the region $\mathcal{R}$ are valid, the other features are discarded. As a result, the image features with low probability of being projection of the landmark are filtered out as physically incorrect.

The shortcoming of this criterion is that more than one image features may fall in the region due to high measurement noise, a lot of measurement artifacts, modeling inaccuracies, or when some image features are in vicinity from each other. The classical approach to this problem is to apply a multiple-hypothesis tracking [125], which forms hypotheses that the image features inside of this region correspond to the landmark and computes probability that these hypotheses are correct. Being in common use for tracking of low-dimensional systems, this approach is hardly applicable to this high-dimensional nonlinear system. The reason is a significant increase of the computational complexity because of simultaneous tracking of all possible hypothesis. This is why we restrict ourselves only to hypotheses that pass the consistency and triangulation criteria.

Consistency Criterion

In the next step, the correspondence function (4.7), which defines the assignment vectors (4.6), is calculated. In this function, among multiple gating criteria the computationally complex triangulation criterion is omitted. Afterwards, it is tested whether the consistency criterion is satisfied.

This criterion, denoted by an indicator function I_C, ensures that no conflicting solutions are returned after the gating. It should be noted that for achieving high accuracy of reconstruction, only the image features that have correspondences in all camera views and that are uniquely assigned to one of the landmarks are considered as measurements.

Triangulation Criterion

When the image features do not uniquely correspond to the landmarks, a triangulation criterion is applied for resolving this conflict. Because of the computational complexity of this criterion, it is checked only on the small set of image features that have not passed the other criteria.

The triangulation criterion selects from all image features, which may be projections of the landmark $\underline{l}_k^i$, only one. Defined by the indicator function

$$
I_T := \begin{cases} 1 & \text{if} \min \sum_{i,j=1}^{n} \left\| \overset{*}{\underline{l}}_k^i - \overset{*}{\underline{l}}_k^j \right\|, \ i > j\,, \\ 0 & \text{otherwise}\,, \end{cases}
$$

this criterion minimizes the Euclidean distance $\|\cdot\|$ between the landmarks reconstructed from the image features in each two camera images, as depicted in Fig. 4.3.

The three-dimensional points $\overset{*}{\underline{l}}_k^i$ and $\overset{*}{\underline{l}}_k^j$ are triangulated by a linear triangulation method proposed in [80] and shortly introduced in Appendix A.

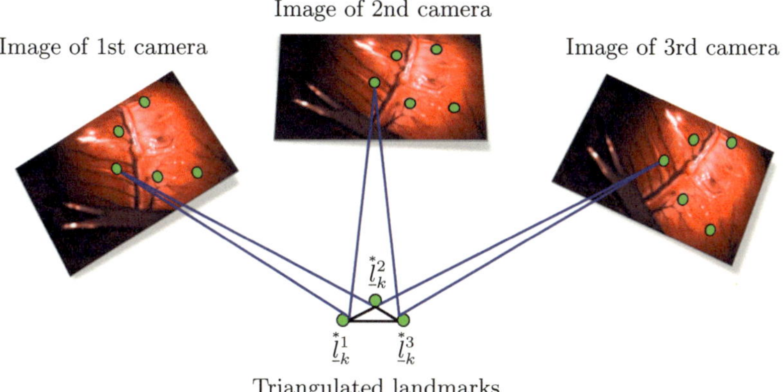

Figure 4.3: Triangulation criterion for detecting the image feature corresponding to a certain landmark. The key idea is to minimize the distance between the landmarks reconstructed based on the image features in each two cameras.

These points are obtained from the pairs of image features $\underline{f}_k^{j,i}$, $j = 1, \ldots, n$ that are considered to be corresponding in each two camera views.

Finally, when the consistency criterion is satisfied, the image features $\underline{f}_k^{j,i}$, $j = 1, \ldots, n$ are assumed to be corresponding to the landmark $\underline{l}_k^i$. Therefore, they are further considered as the projections of the same landmark onto image planes of all cameras.

4.4 Motion Reconstruction

In this section, the heart motion is reconstructed at every time step using nonlinear stochastic estimation that is based upon the physics-based state-space model derived in Chapter 3. By processing the measurement information, the estimates of the current positions of the heart landmarks are obtained, as shown in Fig. 4.4.

The specialty of the estimation is that the stochastic and systematic model and measurement errors are considered. Generally, there are two possibilities for handling these types of errors.

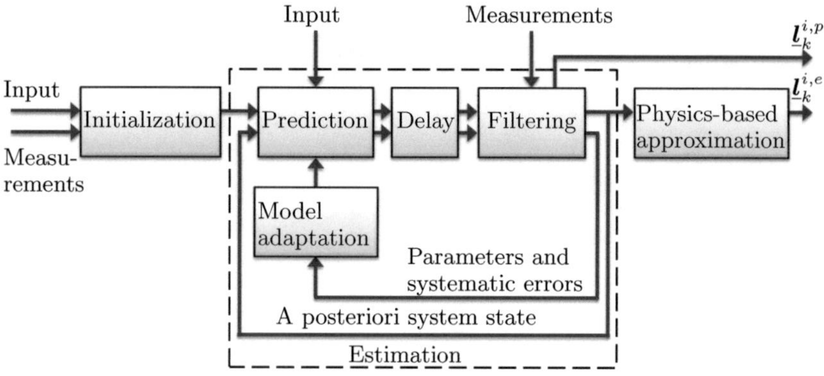

Figure 4.4: Estimation of the a priori $\underline{l}_k^{i,p}$ and a posteriori $\underline{l}_k^{i,e}$ landmark positions.

The common way consists in modeling of stochastic and systematic uncertainties by random quantities with an imposed probability distribution. The downside is that the estimation of the systematic uncertainties can lead to complicated models. Accordingly, due to augmentation of the system state with uncertain quantities, the linear models may become high-dimensional and nonlinear [183].

Another way is to employ imprecise probabilities [220] instead of unique probabilities for characterizing the uncertainties. As it is described in [147, 154, 155], for this purpose sets of densities are used. Here, the systematic errors are characterized by set-valued representations [113, 185], where the stochastic errors are still described by assigned probability distribution. It should be noted that there are no probabilistic relations between the elements inside of the set-valued representations. With respect to the beating heart surgery, consideration of systematic and stochastic errors by means of imprecise probabilities is proposed in our work [241]. In this context, the major benefit of such handling of uncertainties lies in achieving accurate uncertainty bounds. However, there are no probable estimates of the systematic errors that can be employed for improving the quality of the underlying models.

In this thesis, the systematic errors and other unknown quantities, e.g., model parameters, are augmented with the system state. This leads to a conditionally linear system introduced in Section 4.4.1. In order to reduce the curse of dimensionality, the augmented system is decomposed into linear and nonlinear substructures by means of Rao-Blackwellization [176], similarly to [32, 183]. After introducing the initialization of this model in Section 4.4.2, the Section 4.4.3 describes the recursive estimation of the system state in two steps, i.e., the prediction step and the filter step, as illustrated in Fig. 4.4. The positions of the landmarks are obtained at every time step by means of physics-based approximation using the estimated system state. The estimated parameters and systematic errors are used for improving the state-space model at every time step.

4.4.1 Augmented State-Space Model

The augmented state-space model proposed in this section combines the physics-based state-space model introduced in Chapter 3 with the models of the parameters and systematic errors. In order to enable the estimation of the systematic errors of the system and measurement models, as well as unknown model parameters, all these quantities are modeled by random variables.

The parameters of the physical model that vary over time, e.g., Poisson ratio ν_k, Young modulus E_k, stiffness of Robin boundary β_k and material density ρ_k, are assembled in the vector

$$\underline{\theta}_k = [E_k, \nu_k, \beta_k, \rho_k]^{\mathrm{T}} \in \mathbb{R}^{n_\theta} . \tag{4.10}$$

The systematic errors $\underline{\tau}_k$ caused by time-discretization, are supposed to be negligible. This assumption is justified due to small time step and small temporal changes in the heart motion between the time steps. In this case, as follows from (3.53), the errors of temporal discretization are very small.

Conscious of the importance of considering the errors of spatial discretization, the methodology of their a priori estimation published in [241] can be applied. However, in this thesis, the a posteriori estimation of these errors is preferable for continuous improvement of the models. These errors will be adaptively reduced, based upon the feedback from the visual motion compensation, as it is proposed in Chapter 6. The reprojection errors $\underline{e}_k$ can be assumed to be constant over time due to small pixel displacements caused by the heart motion. These errors are determined by the distance between the projections of landmark positions reconstructed by the triangulation method and corresponding to these landmarks image features extracted from the image.

State Augmentation

For estimating the unknown model input $\underline{u}_k$ and time-varying physical parameters $\underline{\theta}_k$ simultaneously with the system state $\underline{z}_k$, these quantities are modeled by random variables, which are augmented in the vector

$$\underline{\xi}_k = \begin{bmatrix} \underline{z}_k \\ \underline{\theta}_k \\ \underline{u}_k \end{bmatrix} = \begin{bmatrix} \underline{x}_k^l \\ \underline{x}_k^n \end{bmatrix} \sim \mathcal{N}\left(\underline{\mu}_k^\xi, \Sigma_k^\xi \right) . \tag{4.11}$$

It should be noted that the augmented state $\underline{\xi}_k$ is decomposed into linear and nonlinear substructures using Rao-Blackwellization [176]. In this way, the efficiency of the stochastic filter used for estimating the augmented system state is improved. The reason for this is that some of the equations evaluated by this filter can be determined analytically instead of computing everything with deterministic sampling. Therefore, as proposed in [32], the augmented state is assumed jointly Gaussian

$$\underline{\mu}_k^\xi = \begin{bmatrix} \underline{\mu}_k^l \\ \underline{\mu}_k^n \end{bmatrix}, \ \Sigma_k^\xi = \begin{bmatrix} \Sigma_k^{l,l} & \Sigma_k^{l,n} \\ \Sigma_k^{n,l} & \Sigma_k^{n,n} \end{bmatrix}$$

with mean $\underline{\mu}_k^\xi$ and covariance Σ_k^ξ.

In the next sections, the state-space model proposed in Chapter 3 is extended by the models of the parameters and systematic errors, which are augmented in a system state. As a result, the augmented state-space model is obtained.

Augmented System Model

For derivation of the augmented system model, first of all, the augmented state is divided into linear and nonlinear substructures

$$\underline{x}_k^l = \begin{bmatrix} \underline{z}_k \\ \underline{u}_k \end{bmatrix}, \ \underline{x}_k^n = \underline{\theta}_k . \tag{4.12}$$

Whereby the nonlinear substructure incorporates the model parameters that act nonlinearly on the augmented system model, the elements of the linear substructure have linear effect on this model. Furthermore, it is supposed that the temporal variations of the physical parameters, so systematic errors are accurately described by a random walk model with additive Gaussian noise according to

$$\underline{u}_{k+1} = \underline{u}_k + \underline{w}_k^u , \ \underline{\theta}_{k+1} = \underline{\theta}_k + \underline{w}_k^\theta . \tag{4.13}$$

Then, similarly to [32], the augmented system model with separate linear and nonlinear dependencies

$$
\underline{\boldsymbol{\xi}}_{k+1} = \underbrace{\begin{pmatrix} \mathbf{B}_k(\underline{\boldsymbol{x}}_k^n)\hat{\underline{u}}_k \\ \underline{0} \\ \underline{\boldsymbol{x}}_k^n \end{pmatrix}}_{\underline{g}_k(\underline{\boldsymbol{x}}_k^n)} + \underbrace{\begin{pmatrix} \mathbf{A}_k(\underline{\boldsymbol{x}}_k^n) & \mathbf{B}_k(\underline{\boldsymbol{x}}_k^n) & \mathbf{0} \\ \mathbf{0} & \mathbf{I} & \mathbf{0} \\ \mathbf{0} & \mathbf{0} & \mathbf{0} \end{pmatrix}}_{\mathbf{G}_k(\underline{\boldsymbol{x}}_k^n)} \underbrace{\begin{bmatrix} \underline{\boldsymbol{z}}_k \\ \underline{\boldsymbol{u}}_k \end{bmatrix}}_{\underline{\boldsymbol{x}}_k^l} + \underbrace{\begin{bmatrix} \underline{\boldsymbol{w}}_k^z \\ \underline{\boldsymbol{w}}_k^u \\ \underline{\boldsymbol{w}}_k^\theta \end{bmatrix}}_{\underline{\boldsymbol{w}}_k^\xi} .
$$

$$(4.14)$$

is derived from (3.54), (4.12) and (4.13). Here, the zero-mean additive system noise is assumed to be white and Gaussian $\underline{\boldsymbol{w}}_k^\xi \sim \mathcal{N}\left(\underline{0}, \boldsymbol{\Sigma}_k^\xi\right)$ with covariance $\boldsymbol{\Sigma}_k^\xi = \mathrm{diag}\left\{\boldsymbol{\Sigma}_k^{w,l}, \boldsymbol{\Sigma}_k^{w,n}\right\}$ assembling the covariance matrices of the linear and nonlinear substructures of the system state.

Augmented Measurement Model

For derivation of the augmented measurement model, the augmented state vector is divided in linear and nonlinear subsections

$$
\underline{\boldsymbol{x}}_k^l = \begin{bmatrix} \underline{\boldsymbol{u}}_k \\ \underline{\boldsymbol{\theta}}_k \end{bmatrix} , \quad \underline{\boldsymbol{x}}_k^n = \begin{bmatrix} \underline{\boldsymbol{z}}_k \end{bmatrix} . \tag{4.15}
$$

Similar to (4.12), the linear substructure of the augmented state incorporates the quantities linear influencing on the augmented measurement model. As a result, this model derived from (3.63), (4.13) and (4.15) can be written for every ith measurement extracted from the image of jth camera in the form

$$
\underline{\hat{y}}_k^{j,i} = \underbrace{\phi^f\left(\mathbf{P}^{j,i}\underline{\varphi}^l\left(\underline{l}^i + \mathbf{H}^i\underline{\boldsymbol{z}}_k\right)\right)}_{\underline{g}_k(\underline{\boldsymbol{x}}_k^n)} + \mathbf{G}_k(\underline{\boldsymbol{x}}_k^n)\underline{\boldsymbol{x}}_k^l + \underline{\boldsymbol{e}}_k^{j,i} + \underline{\boldsymbol{v}}_k^{j,i} , \tag{4.16}
$$

wherein linear and nonlinear dependencies are separated. The matrix $\mathbf{G}_k(\underline{\boldsymbol{x}}_k^n) = \mathbf{0}$ denotes in this equation the zero matrix.

4.4.2   Tracking Initialization

One of the advantages of the physics-based tracking is that a priori knowledge about the motion of the object is incorporated in the estimation process. On the one hand, as will be shown in Chapter 7, this enables bridging of long-term partial and even total occlusions, which is hardly possible with methods that do not exploit any a priori information. On the other hand, the extraction of this information from available data poses a significant challenge of extracting a priori data based solely upon available camera measurements. This section deals with generating a priori knowledge for a proper initialization of the tracking.

It should be noted that, with respect to beating heart surgery, the a priori knowledge, e.g., physical parameters of the heart and configuration of the intervention area, can be obtained from preoperative planning of the beating heart operation. However, the problem arising here is that the heart geometry, as well as the parameters, can change by intraoperative interventions due to the changing environment. Furthermore, sometimes, e.g., in cases of emergency, the preoperative data are not available. In order to cope with this, before tracking of the heart motion, the model parameters and the model configuration are automatically determined by an initialization, as illustrated in Fig. 4.4. Subsequently, the model parameters and systematic errors are continuously estimated by a stochastic filter during the functionality of the system.

Geometry of Physical Model

In order to define the model geometry, it should be remembered that the physical model introduced in Chapter 3 approximates the behavior of the heart wall inside intervention area that is bounded by a mechanical stabilizer. This model with initial configuration representing the heart wall in the mechanical equilibrium, so called undeformed state, is volumetric and takes into account the thickness of the heart wall. Furthermore, it is worth mentioning that for deducing the state-space model, the physical model is discretized by a set of model nodes $\mathcal{M} \subset \overline{\Omega}$ in the spatial domain $\overline{\Omega} = \Omega \bigcup \Gamma_N \bigcup \Gamma_R$, which consists of the open interior Ω, Neumann Γ_N and Robin Γ_R boundaries.

For representing the thickness of the heart wall, the discrete physical model must consist of at least two layers of model nodes. The nodes on the lower surface of the model belong to the Neumann boundary Γ_N, which is subjected to the pressure forces inside of the cardiac chamber. The nodes on the upper surface of the model are visible in all camera views and belong to the interior Ω. Since the continuous motion of the heart hinders the detection of the mechanical equilibrium of the heart wall, the positions of the upper model nodes in the undeformed configuration of the model are defined by positions of the landmarks averaged over a certain time interval, e.g., heart period. Then, the nodes on the Neumann boundary are positioned under the upper model nodes in the direction of the surface normals.

The form of the discrete physical model is defined by a convex hull of a set of image features that has correspondences in all camera views over the whole time interval. This convex hull is a polygon with the smallest possible area that encloses the projections of visible model nodes in every camera view.

Model Parameters

In order to determine the initial values of the model parameters, the response of the physics-based state-space model is adjusted to the displacement of the heart landmarks reconstructed by a triangulation. For this purpose, the constrained nonlinear optimization problem is formulated

$$\underline{\theta} = \underset{\underline{\theta} \in \mathbb{R}^{n_\theta}}{\arg\min} \left(\sum_{k=1}^{K} \left(\underline{\mu}_k^{l^p} (\underline{\theta}) - \overset{*}{\underline{l}}_k \right)^2 \right) \quad \text{such that} \quad \underline{b}^u \leq \underline{\theta} \leq \underline{b}^l \, ,$$

where the vector $\underline{\mu}_k^{l^p}$ collects the mean values of the predicted positions of the landmarks at every time step and the vector $\overset{*}{\underline{l}}_k$ assembles the positions of the landmarks found by a linear triangulation method proposed in [80]. It should be noted that only those image features that are detected in all three cameras are used by a triangulation. The vectors $\underline{b}^u$ and $\underline{b}^l$ incorporate the upper and lower bounds of the model parameters. The vector $\underline{\mu}_k^{l^p}$ is determined by substituting in (3.57) the mean value of the system state

$$\underline{\mu}_k^{z^p} (\underline{\theta}) = \mathbf{A}_k (\underline{\theta}) \underline{\mu}_{k-1}^{z^p} + \mathbf{B}_k (\underline{\theta}) \hat{\underline{u}}_k$$

that is predicted based upon system model (3.54). This results in

$$\underline{\mu}_k^{l^p} = \underline{l} + \mathbf{H} \underline{\mu}_k^{z^p} (\underline{\theta}) \, ,$$

where the matrix

$$\mathbf{H} = \begin{bmatrix} \mathbf{H}^1 \\ \mathbf{H}^2 \\ \vdots \\ \mathbf{H}^N \end{bmatrix}$$

is assembled from individual matrices $\mathbf{H}^i$ defined in (3.64). The system matrix $\mathbf{A}_k$ and the input matrix $\mathbf{B}_k$ are given in (3.51). The mean value of the initial state is computed according to (3.42), wherein the initial values of the displacement and velocity fields are determined by the displacements and velocities of the landmarks $\overset{*}{\underline{l}}_k$ at the initial time step.

To determine the minimum of the optimization problem, a global search iterative algorithm proposed in [215] is used. After a random selection of the initial values of the parameters that do not violate the constraints in the best case, this algorithm consequently examines all trial values. These values are generated by the scatter search algorithm [74]. It should be noted that in order to find the minimum at every iteration, an interior-point algorithm [39], which is most suitable for large and sparse minimization problems, is used. Furthermore, because of the high complex dependencies between the system parameters, the Hessian matrix is approximated by finite differences.

Collection of Measurement Data

Naturally, for initialization of the model, measurement data should be collected over a certain time horizon in order to enable the triangulation of the landmarks, the positions of which are collected in the vector $\overset{*}{\underline{l}}_k$. For this purpose, the motion of the landmarks must be traced using standard approaches that do not incorporate any a priori knowledge about the object behavior.

In this thesis, the positions of the landmarks are reconstructed at every time step using the triangulation method proposed in [80] for initialization of the physics-based tracking. To enable this triangulation, the correspondences between image features in the images acquired at the same time step are established using epipolar, triangulation and consistency criteria, introduced in Section 4.3.2. Instead of the physics-based criterion, the image features that are in the closest neighborhood of the image feature detected at the previous time step, are considered as potential correspondences of this feature. As the applied criteria are sensitive to calibration errors and illumination, false correspondences between image features can occur. For filtering out these artifacts, we exploit the fact that landmarks move continuously. Therefore, the tracks of landmarks with jumps in the trajectory are excluded from the initialization. As a result, the number of the model nodes used for the initialization of the model can be lower than the number of the landmarks.

For initialization of the lost landmarks during the operation of the system, the above-mentioned tracking runs in parallel to the physics-based tracking. It attempts to track the landmarks that are not integrated in the physics-based tracking over a certain time horizon. Once the positions of these landmarks in the undeformed configuration of the physical model are obtained, the landmarks are integrated in the physics-based tracking and their measurements are used by the estimation. In this way, landmarks lost during initialization, as well as landmarks added by a surgeon during functionality of the system, can easily be incorporated in the physics-based tracking at any time.

Initial System State

The initial values of the system state are provided by (3.42), which exploits the displacements and velocities of the landmarks triangulated from the images acquired at the first two time steps after the initialization.

4.4.3 Motion Prediction and Estimation

This section deals with the estimation of the landmark positions. For this purpose, as illustrated in Fig. 4.4, the nonlinear stochastic estimation is used for estimating the augmented state at every time step. Consequently, for estimating the current positions of the landmarks, the state is processed by a physics-based approximation, as illustrated in Fig. 4.4.

Generally, for estimation of the augmented state an arbitrary nonlinear estimator, such as, e.g., unscented Kalman filter [99] or extended Kalman filter [195], can be applied. The sample-based filters are preferred in this thesis, while the extended Kalman filter may introduce large errors in the estimation of the state densities. The reason for this is the first-order linearization of the nonlinear system, which is used for propagating analytically the system state approximated by the Gaussian random variable. The sample-based filters, such as the unscented Kalman filter [99] or the Gaussian filter [89], also approximate the system state by a Gaussian random variable, however, they capture the distribution of this state by a set of sampling points. These points are then propagated through the nonlinear system. In contrast to the extended Kalman filter, which achieves only first-order accuracy, these filters capture the moments of the system state with higher-order accuracy.

In this thesis, a Gaussian filter [89] that exploits the state decomposition on linear and nonlinear substructures as proposed in [32] is applied. In contrast to unscented Kalman filter [99], the Gaussian filter makes use of systematic deterministic sampling that, according to authors [89], leads to a higher accuracy of the moment's approximation.

Predicted Positions of Landmarks

The predicted position of each landmark $\underline{l}_k^{i,p}$ is described according to physics-based approximation (4.2) by the density

$$\underline{l}_k^{i,p} \sim f_k^p\left(\underline{l}_k^i\right) \approx \mathcal{N}\left(\underline{\mu}_{k+1}^{l^{i,p}}, \Sigma_{k+1}^{l^{i,p}}\right) , \tag{4.17}$$

with the mean and covariance

$$\underline{\mu}_k^{l^{i,p}} = \underline{l}^i + \mathbf{\Phi}^i\left(\underline{l}^i,\underline{m}\right)\underline{\mu}_k^{c^p} , \quad \Sigma_k^{l^{i,p}} = \mathbf{\Phi}^i\left(\underline{l}^i,\underline{m}\right)\Sigma_k^{c^p}\left(\mathbf{\Phi}^i\left(\underline{l}^i,\underline{m}\right)\right)^{\mathrm{T}} .$$

Here, the value of the matrix function $\mathbf{\Phi}^i$ is the matrix of shape functions

$$\mathbf{\Phi}^i = \begin{bmatrix} \mathbf{\Phi}^{i1} & \mathbf{\Phi}^{i2} & \ldots \mathbf{\Phi}^{iN_M} \end{bmatrix} , \tag{4.18}$$

the components $\mathbf{\Phi}^{ij}$ of which are defined in (3.44). Futhermore, the vector $\underline{\mu}_k^{c^p}$ denoting the mean of the predicted nodal coefficients is the part of the vector $\underline{\mu}_k^{\xi^p}$, which represents the mean of the predicted augmented system state. The matrix $\Sigma_k^{c^p}$ stands for the covariance of the predicted nodal values. This matrix is the part of the covariance matrix $\Sigma_k^{\xi^p}$ of the augmented system state.

The a priori density of this state

$$\underline{\xi}_k^p \sim f_k^p(\underline{\xi}_k) \approx \mathcal{N}\left(\underline{\mu}_k^{\xi^p}, \Sigma_k^{\xi^p}\right) ,$$

characterized by mean $\underline{\mu}_k^{\xi^p}$ and covariance $\Sigma_k^{\xi^p}$ is provided by a prediction step of the Gaussian filter introduced in Appendix B. It is approximated according to (B.5) by processing the previous augmented system state.

Estimated Positions of Landmarks

The estimated position of each landmark is described according to physics-based approximation (4.2) by the a posteriori density

$$\underline{l}_k^{i,e} \sim f_k^e\left(\underline{l}_k^i\right) \approx \mathcal{N}\left(\underline{\mu}_k^{l^{i,e}}, \mathbf{\Sigma}_k^{l^{i,e}}\right),$$

with mean and covariance

$$\underline{\mu}_k^{l^{i,e}} = \underline{l}^i + \mathbf{\Phi}^i\left(\underline{l}^i, \underline{m}\right)\underline{\mu}_k^{c^e}, \quad \mathbf{\Sigma}_k^{l^{i,e}} = \mathbf{\Phi}^i\left(\underline{l}^i, \underline{m}\right)\mathbf{\Sigma}_k^{c^e}\left(\mathbf{\Phi}^i\left(\underline{l}^i, \underline{m}\right)\right)^{\mathrm{T}},$$

however, in this case, the a posteriori density of the augmented state

$$\underline{\boldsymbol{\xi}}_k^e \sim f_k^e\left(\underline{\boldsymbol{\xi}}_k\right) \approx \mathcal{N}\left(\underline{\mu}_k^{\xi^e}, \mathbf{\Sigma}_k^{\xi^e}\right)$$

is processed. Accordingly, the vector $\underline{\mu}_k^{c^e}$ collecting the mean values of the estimated nodal values is the part of the vector $\underline{\mu}_k^{\xi^e}$, which denotes the mean of this density. The matrix $\mathbf{\Sigma}_k^{c^e}$ stands for the covariance of the estimated nodal values and is a part of the covariance matrix $\mathbf{\Sigma}_k^{\xi^e}$ of the augmented system state.

The a posteriori density of the augmented system state is estimated by processing the incoming camera measurements, ordered in the vector $\underline{\hat{y}}_k$, in the filter step of the Gaussian filter. It is characterized by the first two statistical moments defined according to [32] by

$$\underline{\mu}_k^{\xi^e} = \underline{\mu}_k^{\xi^p} + \mathbf{\Sigma}_k^{\xi^p,y}\left(\mathbf{\Sigma}_k^{y,y}\right)^{-1}\left(\underline{\hat{y}}_k - \underline{\mu}_k^{y}\right),$$

$$\mathbf{\Sigma}_k^{\xi^e} = \mathbf{\Sigma}_k^{\xi^p} - \mathbf{\Sigma}_k^{\xi^p,y}\left(\mathbf{\Sigma}_k^{y,y}\right)^{-1}\left(\mathbf{\Sigma}_k^{\xi^p,y}\right)^{\mathrm{T}}.$$

Here, the vector $\underline{y}_k$ collects all predicted measurements, those mean $\underline{\mu}_k^{y}$ and covariance $\mathbf{\Sigma}_k^{y,y}$ are calculated according to (B.5). The cross covariance matrix $\mathbf{\Sigma}_k^{\xi,y}$ is provided by (B.6).

The estimated model parameters $\underline{\boldsymbol{\theta}}_k^e$ and systematic errors $\underline{u}_k^e$, which are incorporated in the augmented system state $\underline{\boldsymbol{\xi}}_k^e$ according to (4.11), are used for adapting the model to the behavior of the object under observation at every time step.

It should be noted that when no measurements are available, e.g., in case of total occlusions, the a priori positions of the landmarks provide information for where the landmarks most probably are situated. In case of partial occlusions, when only some of the image features can be detected, the positions of these features satisfying the correspondence function introduced in Section 4.3.2 are processed by a Gaussian filter as measurements in the filter step. There, in spite of the fact that not all landmarks are measured, the states of all model nodes are updated because of the correlations between the model nodes.

4.5 Summary

The proposed approach for three-dimensional tracking of the landmarks on the heart surface is of physics-based and probabilistic nature.

Contributions Instead of direct processing the images, this approach takes into account that the images contain the information about the motion of the deformable object. Although tracking of the landmarks is the main issue of this chapter, thanks to the volumetric physical model, which is incorporated in the tracking approach, the state of the entire object can be estimated by using the proposed techniques. In contrast to existing methods for tracking the heart motion, this allows us to get inside the motion of the heart interior by solely processing the camera data. This ability of the tracking approach will be illustrated in Chapter 7, which presents the validation results. The benefit of this is the extension of surgeon's capabilities in performing surgical manipulations during an operation on a beating heart.

A further special feature of the proposed method is the derivation of all equations, such as the correspondence function and the augmented system model used by a stochastic estimation, from the physical model. In this way, a physically reasonable estimation of the object's position is assured. This is especially essential in the case of complete loss of measurement information or highly inaccurate measurements. Furthermore, the generic gating procedure for extracting the measurement information from camera data becomes physics-based due to processing physically reasonable estimates. Formulated by a squared Mahalanobis distance, this physics-based criterion exploits the a priori density of the landmark positions for defining the area of high probability for which the noisy camera measurements are projections for a certain landmark.

In contrast to the broadly used deterministic tracking approaches, the tracking problem is formulated as the estimation of the landmark positions at every time step based upon the previous system state and incoming noisy camera measurements. Such a probabilistic formulation allows consideration of measurement and model uncertainties. This makes the tracking robust to measurement inaccuracies. The information about the uncertainties of the estimated landmark positions, granted by the estimation, is especially important for the safety of the beating heart operation. For example, it can be exploited for definition of virtual features in robotic control, which may prevent the manipulator from entering into hazardous regions of the heart or assist the surgeon in moving the manipulator along desired paths or the heart surface [1, 2, 190].

It is important to note that beyond the stochastic errors, also the systematic errors, such as unknown heart excitation, are determined by the stochastic estimation. For estimating the systematic errors and unknown model parameters simultaneously with the system state, an augmented state-space model is derived, which combines the physics-based state-space model with the random walk models of the parameters. Since this model is conditionally linear, the augmented state is decomposed into linear and nonlinear substructures by means of Rao-Blackwellization in order to achieve a high computational efficiency. The density of the augmented system state is estimated using nonlinear stochastic filter.

Further Developments Future work is devoted to generalize the approach for tracking of the image features that correspond to more than one landmarks. Currently, these image features are discarded, reducing the number of measurements used for estimating the object's motion. In case of large number of measurement artifacts or when image features are located very close to each other, this may lead to total loss of measurement information. A basic approach towards the solution of this problem is to use multiple-hypotheses tracking [20, 125]. By forming association hypotheses, which assign the image feature to the landmark, this approach determines the probability that these hypotheses are correct. Here, special care must be taken of increasing computational complexity due to simultaneous tracking of several hypothesis. An efficient implementation of the algorithms and using simultaneously multiple processing modalities for computation of the hypotheses seem to be unavoidable.

Another issue is the extension of the observability of the system by using additional sensors. A spectral analysis of the beating heart motion shows frequencies up to 20 Hz, as introduced in [45]. Since the heart motion is not perfectly periodic, a high acquisition speed is necessary for sufficiently capturing all frequency components of the heart motion. However, due to the fact that frame rate of the camera achieves usually up to 30 Hz, the motion of the heart cannot be resolved at this scale. Therefore, the manipulator will not be able to cancel the high frequent components of motion by exploiting the predicted data. To deal with this problem, high speed cameras can be used [173], the frame rate of which may achieve 125 Hz. Another possibility is the information fusion of camera data with other sensors. For example, acceleration sensors would significantly improve the ability of the proposed tracking to reproduce the high-frequent motion of the heart. As illustrated in [82], they can resolve the heart motion up to 50 Hz. Furthermore, they are insensitive to contaminations and occlusions, in contrast to cameras.

5 Physics-Based Visual Motion Compensation

This chapter proposes a novel method for visual motion compensation. Its objective is to compensate the changes in the scene caused by the motion of an elastically deformable object observed by a camera system. While this object will be represented as motionless, the changes in the scene that do not originate from the object, e.g., through different coloring, remain visible.

Although the visual motion compensation is essential for different applications, such as, e.g., automated handling of elastic objects like an insertion of a flexible beam into a hole [149] or serving in presence of non-rigid motion [111, 228], this chapter concentrates on computer-assisted surgical operations on a beating heart [150]. As introduced in Section 1.1, here, the visual motion compensation is applied for representing the area of the heart as motionless in order to give a surgeon an impression of operating on a non-moving heart.

The main specialty of the proposed method is that it is physics-based. In contrast to common non-model-based [198] as well as geometric methods for visual motion compensation that directly process the camera images, e.g., geometric warping [73] or morphing [226], this method operates in the physical space. By taking into account that the deformable object underlies the camera images, it preserves the depth information, which is normally lost by other methods due to processing of object's projection on camera images.

The core of this method builds the physics-based image transformation function, which determines the shift of the pixels between two time steps by the displacement of the object. For this purpose, the complete surface of the object is reconstructed at every time step in a physically reasonable fashion by exploiting techniques proposed in the previous chapter. In contrast to approaches [8, 73, 226, 235] commonly used in image processing, this method copes with the measurement and modeling uncertainties, yielding a high accuracy of motion compensation. Thanks to the physical

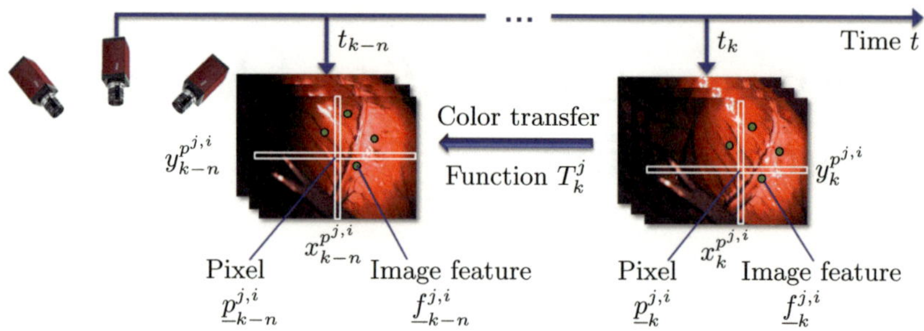

Figure 5.1: Visual motion compensation is formulated as a transformation of the image sequences. This transformation is described by the image transformation function T_k^j, which defines the shift of the pixels in the reference image to the pixels in the current image based upon the known positions of the image features.

model, incorporated in this method, also image regions where color information is lost, e.g., due to occlusions, can be restored. These qualities of the method will be demonstrated in Chapter 7 by evaluation results.

The emphasis of this chapter is on the description of the theoretical foundation of the proposed method. Hence, it starts with Section 5.1, where the visual motion compensation is formulated as a problem of transformation of the image sequences. Then, Section 5.2 gives an overview of the proposed method by introducing the main components of this method. Subsequently, the image transformation function is presented in Section 5.3 and its physical foundation is explained. Consequently, in Section 5.4, the main parts of the visual motion compensation are described in detail. In providing a summary of the most important aspects and an outlook on future work, Section 5.5 concludes this chapter.

The fundamentals of the proposed method were published in [237] in the context of its application in a beating heart surgery.

5.1   Problem Formulation

This section formulates the visual motion compensation as a transformation of the image sequences provided by a camera into the stabilized image sequences. After introducing the definitions that are necessary for understanding the proposed method, it deals with the problem of formation of the stabilized image sequence and its challenges.

Preliminaries Regarding the definitions of the image sequences, it should be mentioned that the stabilized image sequence represents the continuously deformable object under observation as non-moving, where such changes of the object such as different coloring or occlusions by an arbitrary object are shown.

Since the object is observed by a trinocular camera system, there are three stabilized image sequences, which are described by the set of unbounded sequences $\mathcal{R}_{1:k} = \left\{ \mathcal{R}_{1:k}^j \right\}_{j=1}^n$. Each image sequence $\mathcal{R}_{1:k}^j = \left\{ \mathcal{R}_1^j, \ldots, \mathcal{R}_k^j \right\}$ incorporates the stabilized images gathered up to time step k. It should be noted that the stabilized images are represented by a sequence of pixels $\mathcal{R}_k^j = \left\{ \underline{r}_k^{j,i} \right\}_{i=1}^{N_{\mathcal{R}_k^j}}$ with assigned colors $I(\underline{r}_k^{j,i})$.

The image sequences provided by a camera system capture the moving object. It is assumed that the changes in the images are caused only by the motion of this object.

For the purposes of clarity, it should be recalled that all image sequences of the camera system are collected in the set of sequences $\mathcal{P}_{1:k} = \left\{ \mathcal{P}_{1:k}^j \right\}_{j=1}^n$. Furthermore, the image sequence of the individual camera is described by an unbounded sequence $\mathcal{P}_{1:k}^j = \left\{ \mathcal{P}_1^j, \ldots, \mathcal{P}_k^j \right\}$, wherein every image is represented by a sequence of pixels $\mathcal{P}_k^j = \left\{ \underline{p}_k^{j,i} \right\}_{i=1}^{N_{\mathcal{P}_k^j}}$ with assigned colors $I(\underline{p}_k^{j,i})$. The individual pixels are identified by row and column indices assembled in the vector $\underline{p}_k^{j,i} = \left[x_k^{p^{j,i}}, y_k^{p^{j,i}} \right]^{\mathrm{T}}$.

It should be noted that in the following, for sake of simplicity, only the image sequence of one camera is considered because the proposed method for visual motion compensation handles the images sequences of all cameras in the same way.

Formation of the Stabilized Image Sequence The visual motion compensation represents a surgeon the stabilized view of the beating heart. For this purpose, the image sequence provided by a camera should be transformed into the stabilized image sequence. This means that each image

$\mathcal{P}_k^j$ of the image sequence $\mathcal{P}_{1:k}^j$ obtained from the camera should be transformed into a reference image, yielding the stabilized image. One of the previous images of the sequence or an arbitrary virtual image can serve as the reference image. In the following, the image acquired by the camera at time step t_{k-n} is chosen as the reference image $\mathcal{P}_{k-n}^j \in \mathcal{P}_{1:k}^j$. This image, represented by a sequence of pixels $\mathcal{P}_{k-n}^j = \left\{ \underline{p}_{k-n}^{j,i} \right\}_{i=1}^{N_{P_{k-n}^j}}$ shows the object under observation in its mechanical equilibrium, i.e., undeformed state.

For obtaining the stabilized image at time step t_k, the current camera image has to be transformed into the reference image as shown in Fig. 5.1. This transformation occurs by assigning the color of each pixel in the current camera image $I\left(\underline{p}_k^{j,i}\right)$ to a corresponding pixel in the reference image $\underline{p}_{k-n}^{j,i}$. In this way, the color of the pixels in the stabilized image $I(\underline{r}_k^{j,i})$ are determined by the color of the pixels in the current camera image $I(\underline{p}_k^{j,i})$. It should be noted that for this color transfer, the correspondences between the pixels of the reference and current images should be established. The pixels are defined as corresponding when they are images of the same point on the surface of the object at appropriate time steps.

One of the ways to define the correspondences between the pixels is to define the image transformation function

$$T_k^j : \mathcal{P}_k^j \times \mathcal{F}_{k-n}^j \times \mathcal{F}_k^j \to \mathcal{P}_{k-n}^j \, .$$

This function describes the shift of pixels $\underline{p}_k^{j,i} \in \mathcal{P}_k^j$ in the current image to the pixels $\underline{p}_{k-n}^{j,i} \in \mathcal{P}_{k-n}^j$ in the reference image based on the known positions of the corresponding image features $\underline{f}_k^{j,i} \in \mathcal{F}_k^j \subseteq \mathcal{P}_k^j$ and $\underline{f}_{k-n}^{j,i} \in \mathcal{F}_{k-n}^j \subseteq \mathcal{P}_{k-n}^j$ in these images. This is schematically illustrated in Fig. 5.1. The positions of the image features are usually obtained by tracking methods or triangulation [80]. As a result, the positions of the pixels in the stabilized image coincide with their positions in the reference image $\mathcal{R}_k^j = \mathcal{P}_{k-n}^j$ when no transformation errors exist.

Challenges of Image Transformation   Commonly, the corresponding image features $\underline{f}_{k-n}^{j,i} \in \mathcal{F}_{k-n}^j$ and $\underline{f}_k^{j,i} \in \mathcal{F}_k^j$ determine unknown parameters of the image transformation function T_k^j, as introduced in [8, 73, 226, 235]. For example, in the above-mentioned methods, the parameters involve only geometric information about the shift of the image features between the reference t_{k-n} and the current t_k time steps. However, the more a priori information introduced in the processing, the higher accuracy can be achieved [42]. Therefore, the incorporation of the physical characteristics of the object under observation in the image transformation function should lead to an increase in the quality of the visual motion compensation. Furthermore, it should be taken into account that the positions of the image features and consequently, the parameters of the image transformation function are not exactly known. The reason for this is that the positions of the image features extracted from camera images are corrupted by stochastic perturbations caused by measurement inaccuracies.

5.2 Method Overview

The key idea of the visual motion compensation, which is presented in this section, is the transformation of the camera images using the information about the position of the object under observation. This is in contrast to standard methods overviewed in Section 1.3, where the image transformation is provided by direct processing of the images.

Physical Foundation of the Image Transformation Function For formulation of the image transformation function, it is considered that the current and reference images represent projections of the object at two different time steps. Therefore, as illustrated in Fig. 5.2, the shift of pixels $\underline{p}_{k-n}^{j,i} \mapsto \underline{p}_k^{j,i}$ that are projections of the same visible point on the surface of the object is determined by the displacement of this point from the position $\underline{s}_{k-n}^{i} \in \mathcal{S}_{k-n}$ to the position $\underline{s}_k^{i} \in \mathcal{S}_k$. Since the physical characteristics of the object, such as, e.g., material density and elasticity, influence the displacement of the object, the shift of the pixels is physically founded. This is the reason that the proposed image transformation is called physics-based.

The first challenge of this transformation lies in the fact that the positions of all surface points $\underline{s}_k^{i} \in \mathcal{S}_k$ are unknown. This is because the information about the motion of the object surface is extracted from camera images consisting of the finite set of pixels. Accordingly, the surface of the object can be reconstructed only on some discrete points. As it is hardly possible to track all pixels in the image sequence, some image features $\underline{f}_k^{j,i} \in \mathcal{F}_k^{j} \subseteq \mathcal{P}_k^{j}$ detected in the images are usually tracked [160, 181]. By

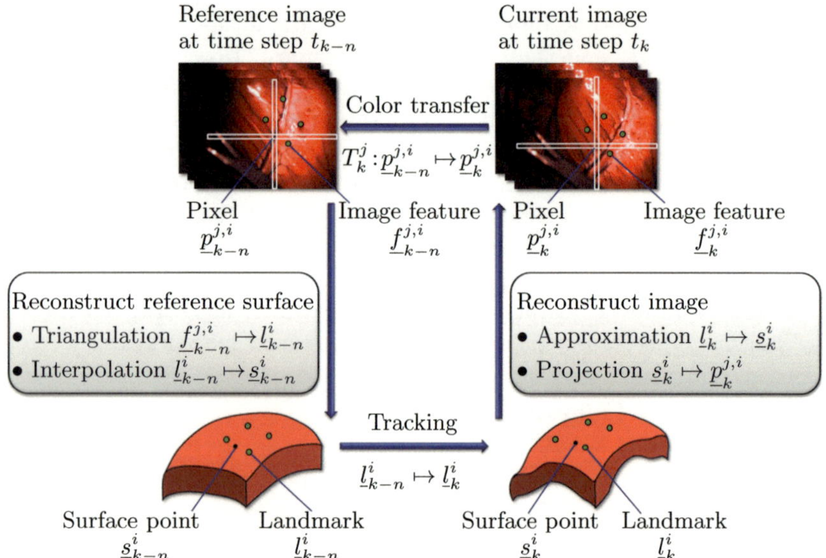

Figure 5.2: An overview of the physics-based visual motion compensation. The shift of the pixels in the reference image to the pixels in the current image is determined by the displacement of the object under observation. This relationship is described by the image transformation function T_k^j.

exploiting the correspondences between these image features, only positions of the landmarks $\underline{l}_k^i \in \mathcal{L}_k \subseteq \mathcal{S}_k$ on the surface of the object can be reconstructed by standard triangulation methods [80]. The problem is further exacerbated by the inaccuracies of the reconstruction arising from camera noise, calibration errors, as well as uncertainties of the image feature extraction.

Three Parts of Physics-Based Visual Motion Compensation In order to cope with these challenges, the physics-based visual motion compensation incorporates three main parts. As schematically illustrated in Fig. 5.2, it includes tracking of the landmarks on the surface of the object, reconstruction of the current camera image that involves the estimation of the current position of the entire object, and surface reconstruction of the object in the undeformed state. The interaction between these parts is listed in Algorithm 2.

Algorithm 2 Visual motion compensation

1: $(\mathcal{L}_{k-n}) \leftarrow \text{Triangulation}\left(\mathcal{F}_{k-n}^{j}\right)$

2: $(\mathcal{S}_{k-n}) \leftarrow \text{SplineInterpolation}(\mathcal{L}_{k-n})$ // Reference surface

3: **for** $t = 1, \ldots, t_k$ **do**

4:     $(\boldsymbol{\mathcal{L}}_k) \leftarrow \text{Tracking}(\mathcal{L}_{k-n})$

5:     $(\boldsymbol{\mathcal{S}}_k) \leftarrow \text{PhysicsBasedApproximation}(\mathcal{S}_{k-n}, \boldsymbol{\mathcal{L}}_k)$ // Current surface

6:     $\left(\mathcal{P}_k^{j}\right) \leftarrow \text{Projection}(\boldsymbol{\mathcal{S}}_k)$ // Current image

7:     $\left(\mathcal{R}_k^{j}\right) \leftarrow \text{ColorTransfer}\left(\mathcal{P}_{k-n}^{j}, \mathcal{P}_k^{j}\right)$ // Stabilized image

8: **end for**

Generally, mapping between the pixels of the reference and current images is obtained by projecting the estimated positions of the physical object onto the image planes of the cameras at respective time steps. Once this mapping is determined, the stabilized images are obtained by color transfer from the pixels in the current images to the pixels in the reference image (line 7).

1) Tracking For tracking of the heart landmarks, the physics-based tracking method proposed in Chapter 4 is used. Based on the physical model, which approximates the behavior of the object, this method estimates the state and parameters of the model, using nonlinear stochastic estimation that processes the image features extracted from incoming camera images. The estimated state provides the information about the positions of the same landmark $\underline{l}_{k-n}^{i} \in \mathcal{L}_{k-n}$ at different time steps (line 4). It is advantageous that the model and measurement uncertainties are considered by estimation. This is why the estimated current position of the landmark $\underline{l}_k^{i} \in \boldsymbol{\mathcal{L}}_k$ is uncertain.

2) Image Reconstruction The image reconstruction yields the estimated positions of all pixels in the current camera images. It should be noted that the positions of these pixels are determined by the current position of the object. Therefore, first of all, the current positions of the surface points $\underline{s}_k^{i} \in \boldsymbol{\mathcal{S}}_k$ are estimated (line 5) by the physics-based approximation introduced in Section 3.4.1. Then, the mean values of these positions are projected in the camera images (line 6), in order to determine the pixels in the current image that originate from these surface points.

2) Reference Surface Reconstruction The reconstruction of the undeformed state of the object is necessary for estimating the current position of this object. It should be noted that, as introduced in Section 3.3.1, the spatial domain of the physical model that approximates the behavior of the object is discretized by a set of model nodes $\mathcal{M} \in \overline{\Omega}$ in order to enable the numerical calculations. The disadvantage in this is certainly the loss of information about the motion of the object between the model nodes. In order to reconstruct this information, the surface of the undeformed model is interpolated (line 2). This interpolation is described in Fig. 5.2 by the mapping $\underline{l}_{k-n}^{i} \mapsto \underline{s}_{k-n}^{i}$, where the model nodes are represented by the landmarks $\mathcal{M} = \mathcal{L}_{k-n}$. The positions of the landmarks in the undeformed configuration of the model are determined by the triangulation of the corresponding image features $\underline{f}_{k-n}^{j,i} \mapsto \underline{l}_{k-n}^{i}$ extracted from the reference images of the camera system (line 1).

5.3 Physics-Based Image Transformation Function

In this section, the mapping between the pixels of the reference and current images is mathematically formulated by the image transformation function. This function is derived by projecting the reference and current positions of the object onto the image plane of the camera.

The image transformation function determines the relationship between the pixels in the current camera image $\underline{p}_{k}^{j,i} \in \mathcal{P}_{k}^{j,i}$ and in the reference image $\underline{p}_{k-n}^{j,i} \in \mathcal{P}_{k-n}^{j,i}$ by the positions of the corresponding surface points at appropriate time steps. For example, when one surface point changes its position from $\underline{s}_{k-n}^{i} \in \mathcal{S}_{k-n}$ to $\underline{s}_{k}^{i} \in \mathcal{S}_{k}$ between the reference and current time steps, then the relationship between the pixels originated from this point is defined by

$$\underline{p}_{k-n}^{j,i} = \underline{\phi}^{f}\left(\underline{\varphi}^{f}\left(\underline{p}_{k}^{j,i}\right) - \mathbf{P}^{j,i}\left(\underline{\varphi}^{l}\left(\underline{s}_{k}^{i}\right) - \underline{\varphi}^{l}\left(\underline{s}_{k-n}^{i}\right)\right)\right) . \qquad (5.1)$$

In this equation, the vector-valued functions $\underline{\varphi}^{f}$ and $\underline{\varphi}^{l}$ introduced in (3.59) and (3.60) describe respective transformations of the image feature or the landmark in homogeneous coordinates. In this context, the function $\underline{\phi}^{f}$ produces inverse mapping, i.e., maps the image feature from the homogeneous coordinates into Cartesian. The projection matrix $\mathbf{P}^{j,i}$ is provided by the camera calibration.

5.4 Transformation of Image Sequences

This section sets forth a detailed description of how the stabilized image is obtained at each time step. For this purpose, the main parts of the physics-based visual motion compensation, such as the tracking of landmarks, the reconstruction of the current image including the estimation of the current position of the object and the reconstruction of the surface of the undeformed object are extensively presented.

Tracking

First of all, it should be noted, that in order to cope with the complex motion of the object, the behavior of the object is described by a physical model introduced in Chapter 3. Then, for tracking of the landmarks on the surface of the object, the physics-based tracking approach introduced in Section 4 is used. It should be recalled that this method propagates the augmented system state $\underline{\boldsymbol{\xi}}_k$ given in (4.11) over time, using the nonlinear estimator proposed in [32, 89]. This state incorporates the system state $\underline{z}_k$, model parameters $\underline{\boldsymbol{\theta}}_k$ and systematic errors $\underline{\boldsymbol{u}}_k$ described by Gaussian random variables. Consequently, the current positions of the landmarks $\underline{l}_k^i \in \mathcal{L}_k$ are estimated by means physics-based approximation (3.57) using the system state $\underline{z}_k$.

Image Reconstruction

Thanks to physical model incorporated in the tracking approach, the current position of the entire object including its surface can be estimated. Therefore, the current camera image can be completely reconstructed by projecting this surface to the image plane of the camera. As a result, the image reconstruction can retrieve the image information that has been lost in the current image, e.g., due to occlusions of the object under observation by other objects. Furthermore, it provides information about which pixels in the current image correspond to which pixels in the reference image.

Current Surface Reconstruction by the Physics-Based Approximation
First of all, the current positions of all surface points $\underline{s}_k^i \in \mathcal{S}_k$ are estimated in the same way as the positions of the landmarks, i.e., by means the physics-based approximation (3.57). It should be noted that these positions are uncertain because of the inaccuracies of the physical model and camera measurements. Their density is determined by plugging into (4.2)

the surface point $\underline{s}_k^i$ instead of the landmark $\underline{l}_k^i$. In this way, the a posteriori position of every surface point is described by Gaussian density

$$\underline{s}_k^{i,e} \sim f_k^e(\underline{s}_k^i) \approx \mathcal{N}\left(\underline{\mu}_k^{s^{i,e}}, \Sigma_k^{s^{i,e}}\right) \tag{5.2}$$

with the mean and the covariance

$$\begin{aligned}
\underline{\mu}_k^{s^{i,e}} &= \underline{s}_{k-n}^i + \Phi^i\left(\underline{s}_{k-n}^i, \underline{l}_{k-n}\right)\underline{\mu}_k^{c^e}\,, \\
\Sigma_k^{s^{i,e}} &= \Phi^i\left(\underline{s}_{k-n}^i, \underline{l}_{k-n}\right)\Sigma_k^{c^e}\left(\Phi^i\left(\underline{s}_{k-n}^i, \underline{l}_{k-n}\right)\right)^{\mathrm{T}}\,,
\end{aligned} \tag{5.3}$$

which depend on the a posteriori system state $\underline{z}_k^e$ incorporating the nodal values $\underline{c}_k^e$ and their time-derivatives. The mean and the covariance of the a posteriori density of the nodal values are denoted by the vector $\underline{\mu}_k^{c^e}$ and matrix $\Sigma_k^{c^e}$ respectively. This density is provided by the filter step of the Gaussian filter introduced in Appendix B. The vector $\underline{l}_{k-n}$ collect the positions of all landmarks of the object in the undeformed state that represent the model nodes. The vector $\underline{s}_{k-n}^i \in \mathcal{S}_{k-n}$ denotes the position of the surface point $\underline{s}_k^i$ of the object in the undeformed state. The matrix Φ^i is determined by (4.18) and (3.44).

When no measurement information is available, the estimated position of the surface point $\underline{s}_k^{i,p}$ depend on the a priori system state. Described by the Gaussian density

$$\underline{s}_k^{i,p} \sim f_k^p(\underline{s}_k^i) \approx \mathcal{N}\left(\underline{\mu}_k^{s^{i,p}}, \Sigma_k^{s^{i,p}}\right)\,, \tag{5.4}$$

they are characterized by the mean and covariance

$$\begin{aligned}
\underline{\mu}_k^{s^{i,p}} &= \underline{s}_{k-n}^i + \Phi^i\left(\underline{s}_{k-n}^i, \underline{l}_{k-n}\right)\underline{\mu}_k^{c^p}\,, \\
\Sigma_k^{s^{i,p}} &= \Phi^i\left(\underline{s}_{k-n}^i, \underline{l}_{k-n}\right)\Sigma_k^{c^p}\left(\Phi^i\left(\underline{s}_{k-n}^i, \underline{l}_{k-n}\right)\right)^{\mathrm{T}}\,.
\end{aligned} \tag{5.5}$$

In this case, the mean vector $\underline{\mu}_k^{c^p}$ and the covariance matrix $\Sigma_k^{c^p}$ describe the a priori density of the nodal values $\underline{c}_k^p$. This density is provided by the prediction step of the Gaussian filter introduced in Appendix B.

As a result, the current position of the entire surface of the object under observation is estimated when no measurement information is available. This is substantial in the context of the beating heart robotic surgery system when the operation area of the heart can be occluded by surgical instruments or blood.

Current Image Formation by the Projection In the next step towards the reconstruction of the current image, the estimated position of the object is projected onto the image plane. For example, when camera measurements are available, the position of the pixel $\underline{p}_k^{j,i} \in \mathcal{P}_k^j$ originated from the surface point $\underline{s}_k^i \in \mathcal{S}_k$ is determined by

$$\underline{p}_k^{j,i} = \underline{\phi}^f \left(\mathbf{P}^{j,i} \varphi^l \left(\underline{\mu}_k^{s^{i,e}} \right) \right) ,$$

where the vector $\underline{\mu}_k^{s^{i,e}}$ represents the mean value of the estimated position of this point. Here, the function $\underline{\phi}^f$ introduced in (3.59) and the function φ^l transform the vectors from homogeneous coordinates into Cartesian and vice versa. The projection matrix $\mathbf{P}^{j,i}$ is provided by the calibration of the jth camera.

It should be noted that since the motion of the surface points $\underline{s}_k^i \in \mathcal{S}_k$ is continuous, their projections $\underline{p}_k^{j,i} \in \mathcal{P}_k^j$ do not necessarily meet the centers of the pixels that are identified by integer indices. Therefore, in order to obtain the color assigned to the surface point in the current image, the bilinear interpolation, e.g., introduced in [76], is used. It determines the color of the pixel, based on the weighted average of the four neighboring pixels. Advantageously, this averaging provides an anti-aliasing effect by producing relatively smooth edges.

Stabilized Image Formation by the Color Transfer For formation of the stabilized image, the color $I\left(\underline{p}_k^{j,i}\right)$ of every pixel in the current image $\underline{p}_k^{j,i} \in \mathcal{P}_k^j$ is assigned to the corresponding surface point $\underline{s}_k^i \in \mathcal{S}_k$. In the next step, these colors are assigned to the pixels in the reference image $\underline{p}_{k-n}^{j,i} \in \mathcal{P}_{k-n}^j$ that correspond to the surface points. In a similar way, the current image can be reconstructed, e.g., in case of occlusions or long time delays between the camera frames. For this purpose, the colors that were assigned to the surface points at a previous time step $I\left(\underline{p}_{k-1}^{j,i}\right)$ are written to the corresponding pixels $\underline{p}_k^{j,i} \in \mathcal{P}_k^j$ in the current image.

Reference Surface Reconstruction

In order to enable the image reconstruction, the undeformed state of the object under observation should be known. This follows from (5.3) and (5.5), where the surface points $\underline{s}^i_{k-n} \in \mathcal{S}_{k-n}$ of the object in the undeformed state are exploited for approximating the current positions of these points.

As the behavior of the object under observation is approximated by the physical model, it should be recalled that, as described in Section 3.3, the undeformed configuration of this model is discretized in space and time to enable the numerical calculations. The downside of this is that the information about the model behavior between the discretization points is lost. Naturally, this information can be reconstructed by the physics-based approximation, according to (5.3) and (5.5), when the positions of the surface points $\underline{s}^i_{k-n} \in \mathcal{S}_{k-n}$ in the undeformed configuration of the model are known.

Since the undeformed configuration of the model is discretized by the set of the landmarks that represent the model nodes, the positions of the surface points are obtained by interpolating between the landmarks $\underline{l}^i_{k-n} \in \mathcal{L}_{k-n}$. For this purpose, the multilevel B-spline interpolation proposed in [117] is used. As stated in [117], the main advantage of this interpolation is that the interpolated surface not only achieves a smooth shape but also closely approximates the scattered points. Performed only once by model initialization, this algorithm generates a bicubic B-spline surface sampled on the grid $\mathcal{G} = \left\{ [k,l]^{\mathrm{T}} \mid 0 \leq k \leq m, 0 \leq l \leq m \right\}$ of data points overlaid on the points $\left[x^{l^i}, y^{l^i} \right]^{\mathrm{T}} \in \mathcal{G}$ defined by the first two coordinates of the landmarks $\underline{l}^i_k \in \mathcal{L}_k$. The value $g_{k,l}$ of every data point is then computed by uniform bicubic B-spline function. As a result, this interpolation yields the set $\mathcal{G}_{k-n} = \left\{ \underline{g}^i_{k-n} \right\}^{N_G}_{i=1}$ of data points $\underline{g}^i_{k-n} = [k,l,g_{kl}]^{\mathrm{T}}$ distributed in the spatial domain $\mathcal{G}_{k-n} \subseteq \overline{\Omega}$ of the undeformed model.

In the next step, in order to determine the correspondences between the surface points of this model and the pixels in the reference image, the points $\underline{g}^i_{k-n} \in \mathcal{G}_{k-n}$ can be projected in the reference image. However, in order to limit the computational complexity, it is suggested to reduce the number of data points $\mathcal{S}_{k-n} \subseteq \mathcal{G}_{k-n}$ by choosing only those points with projections closest to the centers of the pixels in the reference image. The positions of these data points are defined by the minimization function

$$\underline{s}^i_{k-n} = \arg\min_{\underline{g}^i_{k-n}} \left(\left\| \underline{\phi}^f \left(\mathbf{P}^{j,i} \underline{\varphi}^l \left(\underline{g}^i_{k-n} \right) \right) - \underline{p}^{j,i}_{k-n} \right\| \right) , \quad i = 1, \ldots, N_G ,$$

where the Euclidean norm on $\mathbb{R}^2$ is denoted by $\|\cdot\|$. In this way, not only the number of the points representing the surface of the object is reduced, but also the correspondences between the pixels $\underline{p}_{k-n}^{j,i} \in \mathcal{P}_{k-n}^{j,i}$ in the reference image and these points are established. The surface points $\underline{s}_{k-n}^{i} \in \mathcal{S}_{k-n}$ can be now processed by physics-based approximation (5.3) and (5.5) for reconstruction of the current images and formation of the stabilized image.

5.5 Summary

The visual motion compensation proposed in this chapter is formulated as the transformation of the image sequences provided by a camera system into the stabilized image sequences. This transformation is defined by the image transformation function. Overall, it consists of the three main parts: reconstruction of the surface of the object in its mechanical equilibrium, tracking of the landmarks over time, and reconstruction of the current image based upon the current positions of the landmarks. In this way, the correspondences between the pixels of the current and reference images are established. Then, by assigning the colors of the pixels in the current images to the appropriate pixels in reference images, the stabilized images are obtained at the current time step.

Contributions In contrast to other image transformation methods that are based on the direct processing of the camera images, the proposed image transformation function incorporates physical information about the motion of the object under observation. By considering the relationship between the camera images and the position of the object under observation, this function determines the shifts of the pixels in the camera images by the displacement of the object. In doing so, it maintains the information about the three-dimensional position of the object, which would be lost in case of direct processing of images. In addition, thanks to the physical model incorporated in the processing of images, which reflects the physical properties of the object, the information lost in the current images, e.g., due to occlusions, can be restored. This will be demonstrated by experimental results in Chapter 7. Furthermore, it will be shown that the proposed method yields higher accuracy than the methods commonly used for image transformation, not least because of the coping with measurement and modeling uncertainties by means of the nonlinear estimation.

Further Developments   The drawback of this method is that the quality of the motion compensation is not evaluated after the image transformation. Therefore, its deterioration in case of inconsistent estimation of the object's position due to, e.g., inappropriate models or their parameters will not be detected and corrected. This makes the feedback from the stabilized image sequence necessary, in order to monitor and improve the quality of the system. For that reason, an adaptation mechanism dealing with the refinement of the incorporated models is introduced in the next chapter.

Furthermore, it is currently assumed that the changes in the images are caused only by the motion of the one object under observation. When these changes are result of the motion of the multiple objects, this demands the detection of all objects and its individual processing. Hence, surgical instruments should be tracked separately from the heart motion. For this purpose basic approaches for tracking of rigid objects [77], such as color-based strategies or methods relying on a geometric model of the instrument, can be applied.

6 Adaptation of Physics-Based Models

In this chapter, a novel method for continuous monitoring and consistent improvement of the quality of the entire system is proposed. It is motivated by the fact that the quality of the tracking and visual stabilization methods strongly depends on the quality of the models used. The reason is that every model approximating the object's behavior is inaccurate. For example, it may simplify the object geometry, material properties or interaction of the object with the environment in order to make the objects behavior to be describable by the mathematical equations. In spite of this, the models used for image transformation are assumed to be exact in existing methods for tracking and visual motion compensation [5, 19, 78, 114, 175, 186, 198], just to mention a few. Since the quality of the models, as well as of tracking and visual stabilization is not monitored in these works, the deterioration of the motion compensation, for example, due to an inaccurate extraction of the image features or inappropriate models, will not be detected and corrected.

The purpose of the proposed method is twofold. First of all, it strives to refine the physical model for achieving a required level of accuracy via consequent adaptivity. The special challenge lies in monitoring the quality of the model. This is impaired by the fact that camera measurements are available only on some measurement points and, therefore, the accuracy of the model among these points is unknown. Secondly, it is intended to constrain the noticeable increase of computational effort due to refinement of the model. This is a reminder of the fact that the conversion of the physical heart wall model to a discrete state-space model leads to a high-dimensional system, which additionally becomes nonlinear when the unknown model parameters are continuously adapted by estimation methods. In this context, the refinement of the model increases the dimensionality of the system and may enforce its nonlinearity. In summary, the purpose of the proposed method is to achieve a sufficient accuracy by a tolerable complexity of the model also in the areas where no any measurement information is available.

In order to meet these challenges, two main ideas build the core of the proposed method for monitoring and improvement of the quality of the entire system:

- Using error feedback from the stabilized image sequence for quantifying the system accuracy also in the areas where no measurement information is available.

- Consequent reduction of the errors of the model in the areas of high inaccuracies regarding two aspects: mesh refinement of the physical model for reducing discretization errors, and refinement of the physical characteristics of the model for reducing the errors of model parameters.

Furthermore, an important aspect in the context of adaptation is that the refinement of the underlying physical model leads directly to the improvement of the quality of the entire system. The main reason for this lies in the fact that all mathematical models used for motion compensation are derived from the physical model, as illustrated in Fig. 2.2. Therefore, by the refinement of the physical model, the discrete state-space model, physics-based correspondence function, error feedback function, and image transformation function are improved.

This chapter introduces in the next section an adaptation strategy for improving the quality of the tracking and visual motion compensation only where necessary, following which, in Section 6.2, the monitoring and quantification of the quality of the system is discussed. Consequently, Section 6.3 deals with the adaptivity of all mathematical models incorporated in the system. Finally, Section 6.4 summarizes the main points of this chapter. It should be noted that the high quality of the proposed method will be demonstrated in the experimental evaluation in Chapter 7.

The method for monitoring and adaptation of the physics-based system proposed in this chapter was published in [236]. However, the detection and quantification of the errors in the stabilized image sequence and explanation of sources of modeling errors are introduced here in a considerably extended way.

6.1 Adaptation Strategy

The key idea of the adaptation strategy is to use the feedback from the stabilized image sequence, which is provided by the visual motion compensation. As schematically illustrated in Fig. 6.1, this feedback is exploited for improving the quality of the system only where necessary.

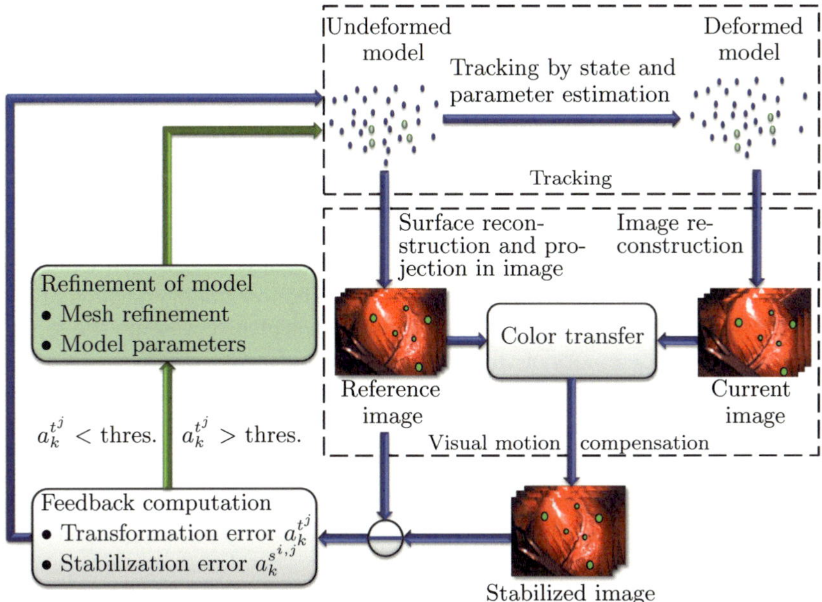

Figure 6.1: Adaptation of the methods based on the feedback from the visual motion compensation. During diagnosis of the system or by the initialization, some of the measurements (small green markers) can be left out from the processing by the estimation and used only for monitoring the quality of the system. The refinement of the spatial discretization of the discrete physical model and its physical characteristics introduces additional degrees of freedom in the state and parameter estimation.

Monitoring of System Quality through the Feedback There are two types of errors that can be extracted from the stabilized image sequence and then used in feedback mechanism: transformation error and stabilization error. Both types of errors are used for evaluating the quality of the entire system. The transformation error is measured in pixels and represents the distance between the stabilized and reference positions of the image features (small green markers). The measured positions of these features are used only for monitoring of the quality of the system. It should be noted that the adaptation is supposed to run during the initialization of the system or as a diagnosis process running parallel to the system. During diagnosis, some of the measurements can be left out from the processing by the estimation and used only for monitoring the quality of the system. Consequently, the transformation error is used as an indicator for the required refinement of the physical model. If this error, averaged over a certain time horizon, is under a predefined threshold, the augmented system state is further propagated by the state and parameter estimation introduced in Section 4.4. In this case, only the model parameters are continuously adapted. However, if the transformation error is above this threshold, the refinement of the physical model is necessary for achieving higher quality of the system. For identifying the low quality regions of the physical model, the stabilization error is used, which represents the difference between the colors of the stabilized and reference images. The main advantage here is that this error provides information about the quality of the system also in the regions where no measurements are available.

Quality Improvement of the System by Models Adaptation For improving the quality of the system in the identified regions, the underlying physical model is refined in two ways. On one hand, the mesh is refined by introducing the additional model nodes (in green) in the initial configuration of the discrete physical model. On the other hand, the physical characteristics of the discrete physical model are improved by assigning model parameters to these nodes. As a result of this refinement, not only the spatial resolution of the discrete physical model is increased, but also the physical characteristics of the model are enhanced with respect to the inhomogeneity of the object's material.

As everything in the proposed system is systematically derived from the physical model, the refinement of the physical model leads to the refinement of all models, e.g., the augmented state-space model used for tracking, or the image transformation function incorporating the estimation of the current position of the entire object and image reconstruction. In this context, the augmented system state is extended by additional state variables and parameters, which describe additional degrees of freedom

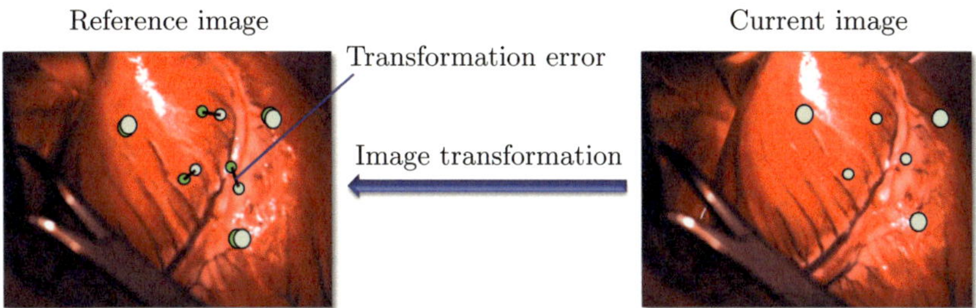

Figure 6.2: Transformation error defined as a Euclidean distance between the reference and stabilized positions of the image features used for monitoring of the quality of the system. This error, measured in pixels, is exploited as an indicator for the required improvement of the quality of the system.

of the state-space model. Consequently, the extended augmented state is determined by the simultaneous state and parameter estimation.

6.2 Feedback from Visual Motion Compensation

This section deals with the quantifying of the quality of the system based upon the stabilized image sequence provided by visual motion compensation. This sequence aims at representing the moving object as motionless. Therefore, the remaining motion in the sequence points out that the deformation of the elastic object under observation could not be completely compensated. For detection and quantification of this motion two types of errors are defined: transformation error and stabilization error. Both errors, used as error feedback for the quality monitoring and improvement, will be presented in this section.

6.2.1 Transformation Error

The transformation error is defined as an Euclidean distance between the reference and stabilized positions of the image features, as presented in Fig. 6.2. Measured in pixels, it is used as an indicator for the required improvement of the quality of the system. When this error, averaged over a certain time horizon, e.g., a period of the heart motion, is over the predefined threshold, the accuracy of the system will be improved by the refinement of the physical model.

Evaluation Points for Quality Monitoring of the System   For monitoring of the quality of the system, all heart landmarks are divided on the evaluation points and measurement points. The current positions of the evaluation points are collected in the set $\mathcal{V}_k = \left\{\underline{v}_k^i\right\}_{i=1}^{N_V}$. The image features that stem from these points, are assembled in the set $\mathcal{Q}_k^j = \left\{\underline{q}_k^{j,i}\right\}_{i=1}^{N_{\mathcal{Q}_k^j}}$. These features, represented in Fig. 6.2 by small green markers, are only used for monitoring of the quality of the system. They are intentionally omitted from the model initialization and stochastic estimation. In contrast, the image features originating from the measurement points form the measurements $\underline{\hat{y}}_k^{j,i} \in \mathcal{Y}_k^j$, which are processed by the Gaussian filter in the filter step. These features are presented in Fig. 6.2 by the large green markers. The current positions of the measurement points are assembled in the set of landmarks $\mathcal{L}_k = \left\{\underline{l}_k^i\right\}_{i=1}^{N_L}$. It should be emphasized that these positions, averaged over a predefined time horizon, identify the set of model nodes that discretize the physical model.

Although the reference and stabilized positions of the image features originated from measurement points commonly almost coincide with each other, as schematically illustrated in Fig. 6.2, it should be noted that high transformation errors may occur on the image features stemming from evaluation points. One of the main reasons for this is that no a priori as well as measurement information about the motion of the object at points corresponding to these features is available in the system.

Definition of the Transformation Error   For a precise definition of the transformation error, it should be recalled that the stabilized image obtained at time step t_k is described by the set of pixels $\mathcal{R}_k^j = \left\{\underline{r}_k^{j,i}\right\}_{i=1}^{N_{\mathcal{R}_k^j}}$ with assigned color $I\left(\underline{r}_k^{j,i}\right)$. The stabilized positions $\underline{\tilde{q}}_k^{j,i} \in \tilde{\mathcal{Q}}_k^j \subseteq \mathcal{R}_k^j$ of the image features $\underline{q}_k^{j,i} \in \mathcal{Q}_k^j$ that originate from the evaluation points $\underline{v}_k^i \in \mathcal{V}_k$, are collected in the set $\tilde{\mathcal{Q}}_k^j = \left\{\underline{\tilde{q}}_k^{j,i}\right\}_{i=1}^{N_{\tilde{\mathcal{Q}}_k^j}}$. Furthermore, it is important to remember that the reference image is described by the set $\mathcal{P}_{k-n}^j = \left\{\underline{p}_{k-n}^{j,i}\right\}_{i=1}^{N_{\mathcal{P}_{k-n}^j}}$ and the color of every pixel in this image is denoted by $I\left(\underline{p}_{k-n}^{j,i}\right)$.

Hence, the transformation error is defined as a Euclidean distance between the reference $\underline{q}_{k-n}^{j,i} \in \mathcal{Q}_{k-n}^j$ and stabilized positions $\underline{\tilde{q}}_{k-n}^{j,i} \in \tilde{\mathcal{Q}}_{k-n}^j$ of the image features corresponding to the evaluation points. In order to determine the quality of the stabilized image at every time step, this error is averaged over the number of the image features $N_{Q_k^j}$ that results in

$$
a_k^{t^j} = \sum_{i=1}^{N_{\tilde{Q}_k^j}} \frac{\left\| \underline{q}_{k-n}^{j,i} - \underline{\tilde{q}}_k^{j,i} \right\|}{N_{\tilde{Q}_k^j}} ,
\tag{6.1}
$$

where the notation $\| \cdot \|$ defines the Euclidean norm on $\mathbb{R}^2$.

One of the key advantages of quality monitoring of the system by exploiting the transformation error is robustness to light changes. The major disadvantage is that the transformation error strongly depends upon the distribution of the image features used for its computation. Therefore, to enable the monitoring of the motion compensation quality in all areas of the heart, the evaluation points must be widely distributed over the heart surface.

6.2.2 Stabilization Error

When the transformation error indicates the required refinement of the model, the stabilization error is computed for getting a deeper insight into the accuracy of the model. This error, measured in colors, is defined as a difference between the stabilized and reference images, as presented in Fig. 6.3. The color of high intensity identify the regions where the model is inaccurate. This information is used for selecting the proper position of the additional model nodes by model refinement.

In order to clarify the reasons for the stabilization errors, it should be recalled that the stabilized image of the scene is obtained by the transfer of the colors of the pixels in the current image to the appropriate positions in the reference image. Therefore, when the positions of the pixels in the current image are accurately estimated, the stabilized image must have exactly the same colors as the reference image. If this is not the case, the models incorporated in the system are inaccurate when it is presumed that no occlusions, e.g., by surgical instruments and changes of light conditions occur.

The stabilization error is defined as a difference between the colors $I\left(\underline{r}_k^{j,i} \right)$ of the pixels in the stabilized image and the colors $I\left(\underline{p}_{k-n}^{j,i} \right)$ of the pixels

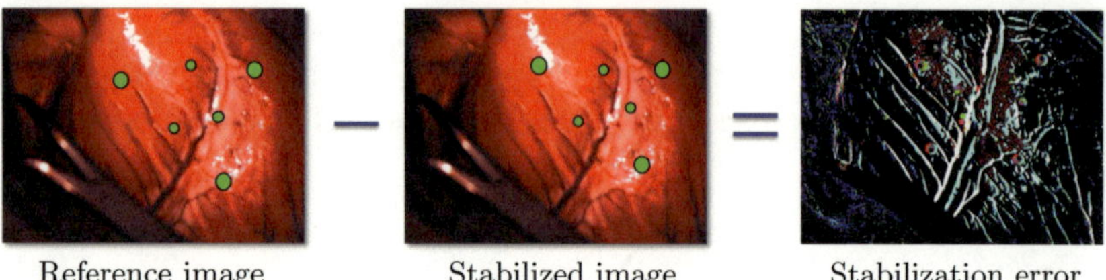

Figure 6.3: Stabilization error as a feedback from visual motion compensation. The sum of these errors over a certain time interval is used for evaluation of the quality of the model in the areas where no measurement information is available.

in the reference image. Then, the quality of the stabilized image sequence involving N_I images collected by the system over a certain time interval is determined by the sum of the stabilization errors computed over all images. For every pixel of the stabilized image, this sum is defined by

$$a_k^{s^{j,i}} = \sum_{i=1}^{N_I} \left(I\left(\underline{p}_{k-n}^{j,i}\right) - I\left(\underline{r}_k^{j,i}\right) \right) . \tag{6.2}$$

It should be noted that this error characterizes the accuracy of the motion compensation in the image plane. In three-dimensional space, it gives the same results, because the surface points have the same colors as the pixels corresponding to them.

Naturally, the stabilization error is strongly sensitive to light changes and illuminations. This makes it difficult to use it as a robust indicator of the quality of the system. Therefore, the changes in the current images that are not caused by the motion of the object under observation, should be filtered out. As a result, this necessitates the detection of occlusions and changes of the object geometry, e.g., due to cutting of heart tissues.

6.3 Adaptation of Models

In this section, the feedback from the visual motion compensation is used for adaptation of the physical model and, therefore, all models incorporated in tracking and visual motion compensation.

Although the modeling uncertainties have been already estimated in Chapter 3, not all modeling errors could be considered. For example, the unknown behavior of the object under observation, or uncertainties of the model geometry, as well as errors of spatial and temporal discretization, cannot be estimated without knowledge of the exact position of the object under observation.

The basic idea of this chapter is to reduce the remaining modeling errors by improving the models in two aspects:

1. Adaptive refinement of the model nodes distribution for tailoring the model mesh to the geometry of the object.

2. Adaptation of the model parameters for adjusting the physical characteristics of the model to the characteristics of the object.

The purpose is to achieve an accurate approximation of the object's behavior by tolerable complexity of the models.

6.3.1 Error Sources of Models

This section deals with an analysis of the main sources of stabilization and transformation errors. The basic idea behind this analysis is to find the connection between these errors and quality of the physical model.

As introduced in Section 6.2, both types of errors occur when the estimated positions of the pixels in the current image are inaccurate. Since these positions are determined by the current position of the object, the above-mentioned errors arise due to divergence between the approximated position of the object and its true position.

It should be recalled that the current position of the object is estimated based upon the physical model, which approximates the object's behavior. Furthermore, this model is converted in a discrete form using discretization methods, i.e., meshless local Petrov-Galerkin mixed collocation method and implicit Euler method, in order to enable the numerical calculations. Consequently, the approximation quality depends on the ability of the discrete model to reproduce the object's motion. For its part, the quality of the discrete model depends on the chosen approximation or discretization parameters, such as polynomial reproduction of the shape functions, size of support domains, the distribution of the model nodes and the approximation of the material characteristics of the object. As the first two

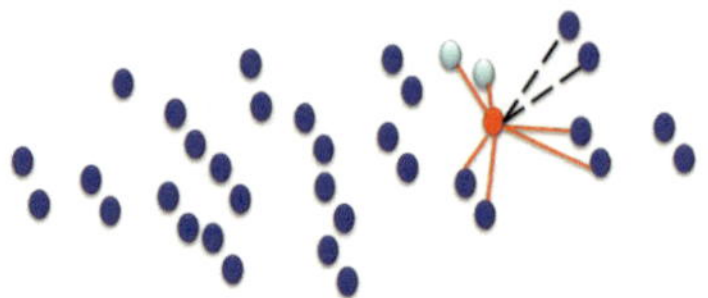

(a) Discretization errors. The local differences in the motion of the surface points (red) among the model nodes (blue) may not be sufficiently resolved when these nodes are at a wide distance from each other.

(b) Errors due to incorrect physical parameters. For considering inhomogeneity of the object material, the additional model parameters are assigned to every inserted model node (green).

Figure 6.4: Main sources of the stabilization and transformation errors. The main reasons for these errors lie in the inaccuracies of the physical model.

parameters are systematically chosen, the latter two are analyzed in the following sections.

Discretization Errors

The coarse distribution of the model nodes can lead to the insufficient spatial resolution of the model. This may result in inaccurate approximation of the motion among the model nodes due to discretization errors.

Sources In order to identify the source of the discretization errors, it should be noted that the spatial resolution of the physical model is determined by its spatial discretization. As a result of element-free spatial discretization, the model is represented by a set of nodes distributed in the spatial domain of the model, as shown in Fig. 6.4(a). Then, the motion of every surface point (red) is approximated, based on a sufficient amount of adjusting model nodes (blue) inside of the support domain of the point (connected with red lines). Therefore, when the model nodes are far from each other, the local differences in the motion of the surface points among these nodes may not be sufficiently resolved. This is especially the case when the large displacements occur or differences in the motion of adjusting surface points are high.

Error Reduction Strategy   For increasing the spatial resolution of the model, the model mesh is refined. This is achieved by inserting additional model nodes (green). In this way, the distance between the model nodes and therefore, the errors of spatial discretization are reduced and as a result, an accurate approximation of the motion also among these nodes can be achieved. However, the insertion of the additional nodes not only contributes to higher spatial resolution of the physical model but also in this way, increases the dimensionality of the system state and therefore, the computational complexity of the estimation of the motion.

Errors of Material Characteristics

The material characteristics of the object are approximated by model parameters with physical meaning, e.g., elasticity modulus or Poisson ratio. Therefore, the accuracy of the physical model and consequently, of the motion estimation, strongly depends upon these parameters.

Sources   Generally, there are two reasons for errors of material characteristics. On one hand, the spatial resolution of material characteristics may be coarse. This is especially the case when the material of the object is strongly inhomogeneous. On the other hand, the model parameters may be incorrect. For example, if the model is stiffer than the object, then, the displacement approximated by this model is smaller then the displacement of the object presuming the same excitation force.

Error Reduction Strategy   For considering the inhomogeneity of the material of the object, the additional model parameters are assigned to every inserted model node (green), as schematically illustrated in Fig. 6.4(b). In this way, the material inhomogeneity among the model nodes is better resolved. The reason for this is that the physical characteristics in the area of every surface points (red) are influenced by model parameters of adjusting model nodes (shaded areas). Therefore, when the model nodes are not widely distanced from each other, the local differences in the material characteristics can be represented more accurately. In the next step, the model parameters are adapted by a state and parameter estimation that processes the incoming camera measurements.

As a result, by introducing additional physical parameters and their adaptation, the errors of material characteristics are reduced. Unfortunately, this occurs at the cost of increasing dimensionality of the system.

Algorithm 3 Adaptive refinement of physics-based models at every time step t_k

---

1:   $(\mathcal{L}_k) \leftarrow \text{Tracking}(\mathcal{L}_{k-n})$

2:   $\mathcal{R}_k^j \leftarrow \text{VisualMotionCompensation}\left(\mathcal{P}_{k-n}^j, \mathcal{P}_k^j\right)$

3:   **if** $a_k^{t^j} >$ threshold **then**

4:      $\check{\mathcal{S}} \leftarrow \text{APosterioriErrorEstimation}\left(\mathcal{P}_{k-n}^j, \mathcal{R}_k^j\right)$

5:      $\check{\mathcal{M}} \leftarrow \text{RefineMesh}(\mathcal{M}, \check{\mathcal{S}})$

6:      $\left(\check{\underline{m}}, \check{\underline{c}}, \check{\underline{r}}, \check{\mathbf{M}}, \check{\mathbf{V}}, \check{\mathbf{K}}, \check{\mathbf{\Phi}}\right) \leftarrow \text{AdaptPhysicalModel}(\underline{m}, \underline{c}, \underline{r}, \mathbf{M}, \mathbf{V}, \mathbf{K}, \mathbf{\Phi})$

7:      $\check{\underline{\theta}}_k \leftarrow \text{RefineParameters}\left(\underline{E}_k^{\check{s}^i}, \underline{\theta}_k\right)$

8:      $\left(\check{\underline{z}}_k, \check{\hat{\underline{u}}}_k, \check{\underline{s}}_k, \check{\underline{w}}_k^z, \check{\mathbf{A}}_k, \check{\mathbf{B}}_k, \check{\mathbf{H}}_k\right) \leftarrow \text{AdaptStateSpaceModel}(\underline{z}_k, \hat{\underline{u}}_k,$
                                           $\underline{s}_k, \underline{w}_k, \mathbf{A}_k, \mathbf{B}_k, \mathbf{H}_k)$

9:      $\left(\check{\underline{\xi}}_k, \check{\underline{g}}_k, \check{\mathbf{G}}_k, \check{\underline{w}}_k^\xi\right) \leftarrow \text{AdaptAugmentedModel}\left(\underline{\xi}_k, \underline{g}_k, \mathbf{G}_k, \underline{w}_k^\xi\right)$

10: **end if**

6.3.2   Adaptive Refinement of Physics-Based Models

This section deals with the adaptive refinement of all physics-based models incorporated in the methods for tracking and visual motion compensation.

Algorithm 3 summarizes the refinement strategy. As the physical model is the basis of all models, the physical model is refined (lines 5-6) at first. This leads, in turn, to the improvement of all models derived from it and used by the tracking and visual motion compensation methods (lines 1-2) and (lines 8-9). The aim of the adaptive refinement is ambivalent: High resolution of the models by tractable dimensionality of the system. In order to achieve this, the models are refined only in the regions where high inaccuracies exist and remain coarse in other regions. The regions of high inaccuracies are provided by the a posteriori error estimation (line 4). This error is described in detail in the following section.

A Posteriori Error Estimation

The error estimation proposed in this section evaluates the difference between the behavior of the physical model and the object under observation with the purpose of finding out the proper positions of additional nodes. In the course of this estimation, the stabilization error, provided by the feedback from visual motion compensation, limits the evaluation to the areas where the visual motion compensation is inaccurate.

The idea of mesh refinement by nodes insertion is inspired by a classical h-refinement strategy in the finite element analysis [33, 103, 116], which is based on a posteriori error estimation. However, a classical a posteriori error estimate considers only numerical errors arising due to spatial discretization, presuming that the available data and models are exactly known. It does not evaluate how sophisticated or how appropriate a discrete mathematical model is for characterizing the behavior of the object under consideration.

In this thesis, the mathematical model formulated by a system of stochastic partial differential equations is not exact and the estimated position of the object is uncertain. With the aim at taking into account these uncertainties, the a posteriori error ϵ_k^i is estimated by the squared Mahalonobis distance [136]. Hence, the divergence between the behavior of the physical model and the object under observation at every surface point $\underline{s}_k^i \in \mathcal{S}_k$ is described by the error feedback function, which is defined by

$$\epsilon_k^i = \left(\underline{\mu}_k^{s^{p,i}} - \underline{\mu}_k^{s^{e,i}} \right)^{\mathrm{T}} \left(\Sigma_k^{s^{p,i}} + \Sigma_k^{s^{e,i}} \right)^{-1} \left(\underline{\mu}_k^{s^{p,i}} - \underline{\mu}_k^{s^{e,i}} \right) < k \qquad (6.3)$$

between the predicted $\underline{s}_k^{p,i} \sim f_k^p \left(\underline{s}_k^i \right) \approx \mathcal{N} \left(\underline{\mu}_{k+1}^{s^{p,i}}, \Sigma_k^{s^{p,i}} \right)$ and estimated $\underline{s}_k^{e,i} \sim f_k^e \left(\underline{s}_k^i \right) \approx \mathcal{N} \left(\underline{\mu}_k^{s^{e,i}}, \Sigma_k^{s^{e,i}} \right)$ positions of this point, approximated by the Gaussian densities, with respective means and covariances. These densities are determined by (5.2) and (5.4). Here, the parameter k ensures, with the certain selected probability $P_k := P(\epsilon_k^i < k)$, sufficient quality of the model for characterization of the object. This parameter is a chi-square distributed random variable with three degrees of freedom. It is determined according to the statistical table of chi-square distribution given, e.g., in [20], by the familiar confidence region and three degrees of freedom.

It should be noted that the evaluation of the error feedback function (6.3) for all surface points will lead to a high computational effort. Therefore, in order to harness the computational complexity, the error is estimated only on the surface points, where the stabilization error is above some preset tolerance. For this purpose, the stabilization error is first converted in gray scale, as illustrated in Fig. 6.5. Then, the obtained image is binarized. In this way, only the pixels, the gray intensity levels of which

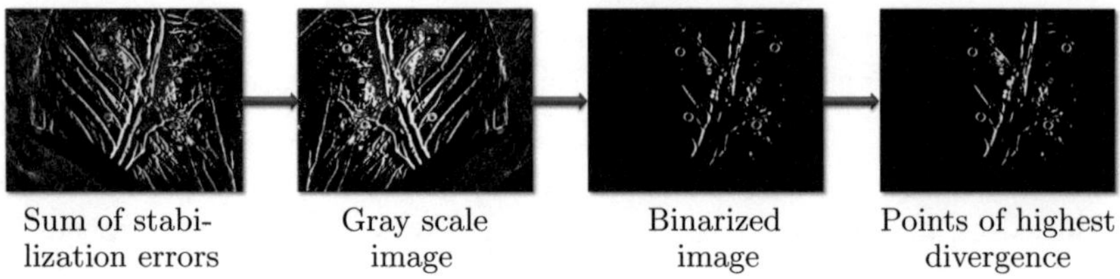

Figure 6.5: Selection of the surface points for refinement of the model discretization. The estimation error is evaluated only on those points, where the stabilization error is above some arbitrarily preset tolerance.

exceed a certain preset tolerance, become white. Finally, from the surface points corresponding to these pixels, only those are selected that violate the requirement $\epsilon_k^i > k$. The violation of this requirement highlights that the quality of the model is insufficient at these points with the relatively high probability of $1 - P_k$.

Refinement of the Physical Model

In this section, the space-discrete physical model set forth in (3.45) is improved regarding **two aspects**: refinement of the spatial discretization and adaptation of model parameters for finer resolution of the material inhomogeneity.

1) Spatial Discretization First of all, in order to reduce the errors of the spatial discretization, additional model nodes $\underline{\breve{s}}^i \in \breve{\mathcal{S}}$ are inserted in the undeformed configuration of the discrete physical model (lines 4-5 of Algorithm 3). These nodes collect the surface points as well as the points placed under the surface points in the direction of the surface normals. In this way, not only the surface but also the interior of the model is refined. As a result, the number of the model nodes increases to $\breve{\mathcal{M}} = \breve{\mathcal{S}} \cup \mathcal{M}$ so that additional degrees of freedom are introduced in the physical model. Hence, additional nodal values $\underline{c}^{\breve{s}^i} \in \mathbb{R}^3$ and their derivatives are assigned to the inserted model nodes $\underline{\breve{s}}^i \in \breve{\mathcal{M}}$.

Consequently, the displacement of the points near the inserted model nodes is still approximated according to (3.24). However, the model nodes in the set of neighboring nodes $\mathcal{V}$ are changed. Now, the widely distanced node is excluded from this set, whereby the newly inserted and less distant model node is included. As a result, the set of neighboring nodes, after insertion of the additional one $\check{\mathcal{V}} \neq \mathcal{V}$, consists of the model nodes in the nearest neighborhood of this point. The number of the neighboring nodes $N_V = N_{\check{V}}$ remains constant. In this way, the approximation of the points' displacement becomes more local. This means that the widely distant points that move in a different way have little influence on each other. The spatial resolution of the model is increased.

The price for this is the enhancement of the dimensions of the space-discrete physical model (3.45), as illustrated in Algorithm 3 (line 6). Hence, the vector $\underline{m} \in \mathbb{R}^{3N_M}$ collecting the positions of the N_M model nodes is enhanced by the positions of the additional model nodes $\underline{\check{s}}^i \in \check{\mathcal{S}}$ collected in the vector $\underline{\check{s}} \in \mathbb{R}^{3N_{\check{s}}}$. Furthermore, the vector of nodal values $\underline{c} \in \mathbb{R}^{3N_M}$ is extended by the coefficients $\underline{c}^{\check{s}^i} \in \mathbb{R}^3$ assigned to the inserted model nodes $\underline{\check{s}}^i \in \check{\mathcal{S}}$. Therefore, the refined space-discrete physical model is determined by plugging in (3.45) the enhanced vectors of model nodes, nodal coefficients, and collocation points

$$\underline{\check{m}} = \begin{bmatrix} \underline{m} \\ \underline{\check{s}} \end{bmatrix} , \quad \underline{\check{c}} = \begin{bmatrix} \underline{c} \\ \underline{c}^{\check{s}} \end{bmatrix} , \quad \underline{\check{r}} = \begin{bmatrix} \underline{r} \\ \underline{\check{s}} \end{bmatrix} . \tag{6.4}$$

Here, the vector $\underline{c}^{\check{s}} \in \mathbb{R}^{3N_{\check{s}}}$ collects the nodal values $\underline{c}^{\check{s}^i} \in \mathbb{R}^3$ of the inserted model nodes $\underline{\check{s}}^i \in \mathbb{R}^{3N_{\check{s}}}$. From this it follows that the global mass matrix $\mathbf{M} \in \mathbb{R}^{3N_R \times 3N_M}$, the damping matrix $\mathbf{V} \in \mathbb{R}^{3N_R \times 3N_M}$, the stiffness matrix $\mathbf{K} \in \mathbb{R}^{3N_R \times 3N_M}$, and the matrix of shape functions $\mathbf{\Phi} \in \mathbb{R}^{3N_R \times 3N_M}$ assembled from local matrices (3.36) similarly to (3.43) should be also extended for considering additional model nodes. Here, N_R denotes the number of collocation points in original model. The matrices of the refined physical model are determined by the horizontal concatenation

$$\check{\mathbf{M}} = \begin{bmatrix} \mathbf{M}, \mathbf{M}^{\check{s}} \end{bmatrix} , \quad \check{\mathbf{V}} = \begin{bmatrix} \mathbf{V}, \mathbf{V}^{\check{s}} \end{bmatrix} , \quad \check{\mathbf{K}} = \begin{bmatrix} \mathbf{K}, \mathbf{K}^{\check{s}} \end{bmatrix} , \quad \check{\mathbf{\Phi}} = \begin{bmatrix} \mathbf{\Phi}, \mathbf{\Phi}^{\check{s}} \end{bmatrix} , \tag{6.5}$$

where in each case the global matrix is combined with the matrix determined by the inserted model nodes $\underline{\check{s}}^i \in \check{\mathcal{S}}$. The elements of this matrix are local matrices computed according to (3.36) for every collocation point $\underline{\check{r}}^i \in \overline{\Omega}$ and inserted model nodes $\underline{\check{s}}^i \in \check{\mathcal{S}}$ substituted in the vector $\underline{m}^i \in \mathcal{M}$.

2) Model Parameters Furthermore, the space-discrete physical model given in (3.45) is adapted to the inhomogeneity of the physical characteristics of the object under observation. This is achieved by describing the elasticity of the heart tissues at inserted model nodes $\breve{\underline{s}}^i \in \breve{\mathcal{S}}$ by additional parameters $\underline{E}_k^{\breve{\underline{s}}^i}$ denoting the elastic modulus of the heart in the area where additional nodes are inserted. This leads to the enhancement of the parameter vector $\underline{\boldsymbol{\theta}}_k$ given in (4.10) (line 7 of Algorithm 3). It should be noted that these parameters are time-dependent because they will be estimated by the stochastic filter at each time step. The material characteristics are now defined at these nodes by the material matrices $\mathbf{C}^j$ computed according to (3.29), where the Lamé constant introduced in (3.7) is determined by the parameters $\underline{E}_k^{\breve{\underline{s}}^i}$. As a result, the extended damping matrix $\breve{\mathbf{V}}$ and the stiffness matrix $\breve{\mathbf{K}}$, which depend on the material matrix $\mathbf{C}^j$ according to (3.36), are also determined by the additional parameters assigned to the inserted model nodes. In this way, the physical characteristics of the model are locally refined.

Finally, it should be noted that, in order to be aware of the non-singularities of the matrices, the same number of collocation points N_R as the number of model nodes $N_{\breve{M}}$ is chosen. The further advantage of this is the finer discretization of the random fields describing the random model input (3.38) and uncertain initial conditions (3.41). These fields are now represented by a higher amount of random variables.

As a result, the discrete physical model becomes a higher spatial resolution due to the insertion of the additional model nodes. Furthermore, the material inhomogeneity of the heart tissues is taken into account. In order to bound the dimensionality of the system, the model is improved only in the local areas where the modeling inaccuracies are detected by the feedback mechanism.

Adaptation of Physics-Based Models

As the physics-based model (3.45) builds the core of the tracking and visual motion compensation, the refinement of this model leads to the improvement of the spatial resolution of all models derived from it.

1) State-Space Model Hence, the state-space model consisting of the physics-based system model (3.54) and the measurement model (3.63) is adjusted for reducing the errors of motion compensation (line 8 of Algorithm 3). After refinement of the physical model, this model is determined by the extended system $\breve{\mathbf{A}}_k \in \mathbb{R}^{6N_{\breve{M}} \times 6N_{\breve{M}}}$, input $\breve{\mathbf{B}}_k \in \mathbb{R}^{6N_{\breve{M}} \times 6N_{\breve{M}}}$, and measurement $\breve{\mathbf{H}}_k \in \mathbb{R}^{6N_L \times 6N_{\breve{M}}}$ matrices that are obtained by plugging the matrices (6.5) into (3.51), (3.64), and (3.58). The extended system state $\underline{\check{z}}_k \in \mathbb{R}^{6N_{\check{M}}}$ is computed according to (3.47), where the extended vector of nodal values defined in (6.4) is exploited. In addition, the vectors describing the modeling systematic errors $\underline{\check{s}}_k \in \mathbb{R}^{6N_{\check{R}}}$ and stochastic errors $\underline{\check{w}}_k^z \in \mathbb{R}^{6N_{\check{R}}}$ are enhanced.

2) Augmented State-Space Model   In a similar way, the augmented state-space model, consisting of the augmented system model (4.14) and the augmented measurement model (4.16), is refined (line 9 of Algorithm 3). It should be noted that in the course of the model refinement, the dimensions of the augmented state $\underline{\check{\boldsymbol{\xi}}}_k$ given in (4.11) are increased. The reasons for this is the higher dimensions of the extended system state $\underline{\check{z}}_k \in \mathbb{R}^{6N_{\check{M}}}$. Furthermore, the unknown model input $\underline{\check{u}}_k \in \mathbb{R}^{6N_{\check{M}}}$ is enhanced by the random variables $\underline{u}^{\check{s}^i}$ assigned to the inserted model nodes $\check{s}^i \in \check{\mathcal{S}}$. Finally, the vector of model parameters (4.10) is extended by the additional parameters $\underline{E}_k^{\check{s}^i}$.

3) Physics-Based Approximation   Consequently, the predicted and estimated positions of the landmarks, as well as the positions of all surface points, are corrected (line 1 of Algorithm 3). The main reason for this lies in the fact that the moments of the Gaussian densities (4.17), (4.17), (5.2), and (5.4) that describe these positions, are derived from physics-based approximation (3.24) with finer discretization. Furthermore, the nodal values of this approximation are determined based on the augmented state-space model with finer discretization. Moreover, this model considers the inhomogeneity of the object's material.

4) Correspondence Function   The adaptation of the predicted positions of the landmarks causes the adjustment of the correspondence function (4.7), which is included in the tracking method (line 1 of Algorithm 3). This function incorporates the physics-based criterion (4.9), which is determined by the predicted measurements. These measurements are estimated by projecting the already correct predicted positions of the landmarks onto the image plane of the camera.

5) Image Transformation Function   Moreover, the physics-based image transformation function (5.1), determined by the estimated positions of the surface points (5.2), is also modified (line 2 of Algorithm 3). This is due to the improvement of the augmented state-space model used by the stochastic estimation and the finer discretization of the physics-based approximation (3.24).

6) Error Feedback Function Finally, the error feedback function (6.3) defined by the squared Mahalanobis distance [136] between the estimated and predicted positions of the surface points is adjusted at the next time step (line 4 of Algorithm 3).

6.4 Summary

This section proposes a novel methodology for continuous monitoring and improvement of the quality of the entire system for tracking and visual motion compensation.

Contributions One of the special things about this adaptation method is that the feedback from the stabilized image sequence is used for the refinement of the models only when and where it is necessary. The transformation error denoting the distance between the stabilized and reference positions of the evaluation points indicates if the quality of the system must be improved.

The further contribution of this chapter is the quality quantification of the system. The stabilization error, which is defined by the color difference between the stabilized and reference images, allows for evaluation and improvement of the quality of the system even in the areas where no measurements are available. This error is used for selecting the proper positions of the additional nodes, aiming at refinement of the model. For this purpose, from all points with high stabilization error are considered only those that have high probabilities of modeling inaccuracies are considered. These points are determined by the error feedback function, which evaluates the squared Mahalanobis distance between the predicted and estimated positions of the object.

One of the sophisticated features of the proposed adaptation strategy is that it makes use of the fact that all physics-based models of the system are derived from one physical model. Therefore, the adjustment of the physical model to the motion of the object under observation causes the improvement of all other models. These models are improved regarding two aspects. On one hand, the spatial resolution is increased by finer discretization in the areas where the modeling inaccuracies are detected. On the other hand, the physical characteristics of the object becomes more differentiated in these areas, yielding a better resolution of the inhomogeneity of the object's material.

Further Developments Considering the fact that the stabilized error is sensitive to the changing light conditions, it may be misinterpreted when the deformable object is occluded by other objects. Therefore, the robust detection of changing light conditions, illuminations, and instruments in the field view of the camera is of paramount importance for ensuring the proper adaptation of the models.

Even if the proposed adaptation method provides high accuracy, as will be shown in Chapter 7, it can be further elaborated. In addition to the errors of spatial discretization and model parameters, the discrepancy of the physical model arising due to its insufficient order can be analyzed and adaptively selected. This could further improve the convergence of the adaptation and lead to higher quality of the motion compensation.

7 Evaluation

This chapter presents the evaluation of the methods for motion compensation. The accuracy of the tracking and of the visual motion compensation is determined by the processing of the stabilized image sequence due to the fact that the motion compensation is as accurate as the motion estimation. Hence, in case of highly accurate motion compensation, the stabilized image representing the moving object as motionless should remain constant over time. The remaining motion in the stabilized image shows inaccuracies in estimation.

With regard to the use of motion compensation in beating heart surgery, both methods are evaluated by experiments on a pressure regulated artificial beating heart that simulates the motion of a mechanically stabilized real heart. One of the points verified by the evaluation is the three-dimensional reconstruction of the heart position in case of partial and total loss of measurement information. The other point to be verified is the capability of the system to compensate the motion of the moving object under observation. In this context, the validity of the image stabilization, as well as of the three-dimensional motion compensation is checked. Furthermore, the system is tested in respect to its sensitivity to the changes of the heart rhythm and varying amount of available measurements.

The reliability of the physics-based motion compensation is compared with the standard method for image warping [8] that exploits the geometric image transformation. The maximum error of the three-dimensional motion compensation provided by this method achieves 0.648 mm. This is in contrast to the maximum error of 0.357 mm, provided by the physics-based motion compensation. Furthermore, the adaptation of the entire system based on the feedback from the stabilized image sequence reduces the maximum error to 0.281 mm. Another special characteristic of the physics-based motion compensation is its ability to restore the current and stabilized images when the image acquisition fails, e.g., due to occlusions. This is of paramount importance for application in a beating heart surgery system where surgical instruments operate in the field of camera view. Additionally, the three-dimensional position of the entire heart wall including its surface and interior is reconstructed – which differentiates the proposed method from approaches commonly used for tracking and motion compensation of the beating heart motion. This reconstruction is possible even when no measurement information is available.

This chapter is structured as follows. Section 7.1 gives an overview of the experimental setup and introduces error measures used for the quantification of the accuracy of the methods. In Section 7.2, the experimental results provided by the tracking approach and physics-based method for visual motion compensation are illustrated and discussed. Finally, this chapter concludes with Section 7.3, where the main points of the evaluation are summarized and intended evaluation of the system in in-vivo experiments is discussed.

In addition to some experimental results published in [236], additional experiments are introduced in order to clarify the capabilities and superior properties of the proposed methods for tracking and visual compensation.

7.1 Experimental Environment

This section describes the experimental environment for testing and verification of the system for motion compensation. After introducing the setup of experiments in Section 7.1.1, Section 7.1.2 presents the evaluation criteria used for error analysis.

7.1.1 Setup

The system settings presented in this section are used in all experiments when others are not mentioned. Although the working environment described here does not directly correspond to one of the beating heart operation, the proposed methods for tracking and visual motion compensation are verified by real experiments on a physical deformable object. At the first step towards the application of these methods in a computer-assisted beating heart surgery, this allows us to testify their quality and show their diverse capabilities before validation of the system in in-vivo experiments.

Artificial Beating Heart as Simulator of Beating Heart Motion The proposed methods for tracking and visual motion compensation are verified by the experiments on a pressure regulated artificial beating heart presented in Fig. 7.1(a). This heart is continuously subjected to the time-varying pressure determined by the predefined pressure signal. In this way, the motion of the mechanically stabilized beating heart is simulated. According to the medical study performed in [43], the remaining motion of the beating heart after mechanical stabilization using the Octopus stabilizer [61] is quantified by an average maximum excursion of 0.596 ± 0.200 mm in x direction, 0.811 ± 0.235 mm in y direction, and 2.6 mm in z direction for the case when the animal has a weight of 65 kg. The measured maximum excursion of the artificial heart achieves 2.6 mm and 4 mm in x- and y-direction, respectively, and 5 mm in z-direction. This motion is caused by the pressure signal with the amplitude of 100 hPa and the frequency of 1.2 Hz. For comparison, according to [43], the motion of the destabilized heart can reach the maximum excursion of 7.2 ± 0.3 mm.

Trinocular Camera System The motion of the artificial heart is captured by a trinocular camera system installed at a distance of 50 cm above the heart, as depicted in Fig 7.1(b). The three cameras PIKE F-210C [4] with image resolution of 1920 pixel $\times$ 1080 pixel achieve an accuracy of 0.2 mm/pixel in y-direction and 0.11 mm/pixel in x-direction of the image coordinate system. The camera baselines are about 57 cm, their focal length is about 35 mm, and the field of view is 12.8 cm $\times$ 17.02 cm. The image size of each camera is reduced by cutting out the defined regions of interest. For calibration of the cameras the multi-camera calibration algorithm proposed in [205] is used. The achieved accuracy of the calibration defined by the maximum reprojection error is 1.2 pixel. The measurement information is transferred via the communication protocol FireWire IEEE 1394b with a frame rate of about 23 Hz.

Camera Measurements It should be noted that for evaluation of the methods, the artificial landmarks are attached to the surface of the heart, as shown in Fig. 7.1(a). The advantage of using these landmarks is the robust and accurate extraction of the measurement information during the beating heart operation. Another advantage is the independence of the system from the object texture. Nevertheless, the attachment of the artificial landmarks can decelerate the begin of the surgical activity. In this context, it is important to note that the proposed methods can be also applicable to the tracking of natural landmarks, since the image features originating from the landmarks represent the input information in the physics-based system.

The camera measurements are determined by the centers of the green circular markers, which denote the projections of the artificial landmarks

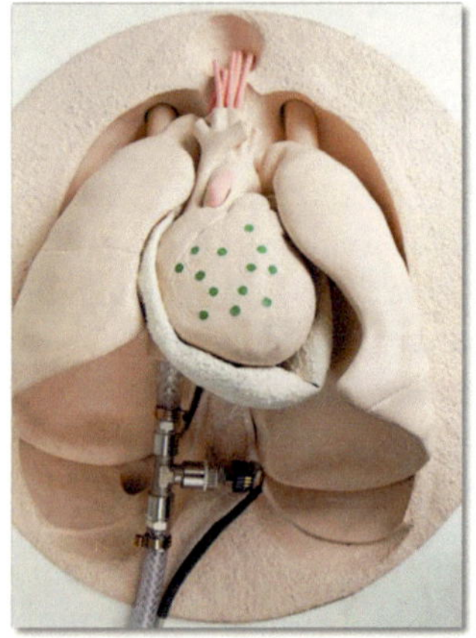
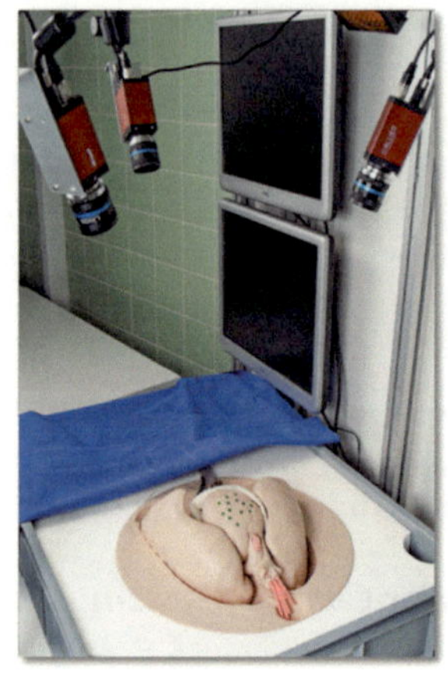

(a) Pressure   regulated   artificial
heart.

(b) The artificial beating heart
is observed by a camera system.

Figure 7.1: Experimental setup for evaluation of the methods.

onto the image plane of the camera. These markers are detected in the camera image by a simple color segmentation algorithm, which introduces some criteria in order to ensure that the image features extracted from the image originate from the markers.

The segmentation algorithm assumes that the markers can be represented only by the image features that lie inside of the hull that includes the shape of the object under observation. Furthermore, when RGB color system is used, the intensity of the color components of these features must be inside of the thresholds of the color channels, i.e., minimal color intensity of the green channel and maximal color intensity of the blue and red channel. Moreover, the image features must be determined by a sufficient amount of pixels.

In order to obtain these features, first, the image is binarized so that only pixels inside the defined color thresholds are converted to white pixels. Consequently, the morphological operation of opening [206] that is obtained by an erosion followed by a dilation of the resulting image is applied. Then, if the obtained convex white zones have enough pixels, a rectangle surrounding every zone is built. The center of this rectangle defines the center of the marker that is supposed to be an image of the artificial landmark.

Real and Simulated Loss of Measurements   The sensitivity of the system to the number of available measurements is evaluated in two ways. On one hand, the loss of measurements is simulated. In this case, although the measured positions of the image features are available, they are not processed by the Gaussian filter in Section 4.4.3. As a result, less measurements are involved in the system than actually present. In this way, the predicted position of the landmark can be compared with its measured position that was assumed to be unknown for the entire system. On the other hand, the capability of the system is also tested by real experiments, whereby the object under observation is occluded by other objects. In this case, no measurement information can be extracted from the occluded regions.

7.1.2   Error Measures

Overall, there are three error measures used for analysis of the motion compensation quality: average transformation error, sum of stabilization errors, and average motion compensation error. The first two errors determine the quality of the visual motion compensation in the image plane. The third error verifies the accuracy of the system in the three-dimensional space.

Similarly to the setup for quality monitoring of the system described in Section 6, all artificial landmarks attached to the heart surface are divided on the evaluation points and measurement points. The image features, represented in Fig. 7.2 by small green markers enumerated by indices in blue, stem from evaluation points. The features originated from measurement points are presented in Fig. 7.2 by the large green markers enumerated by indices in black.

For evaluating the accuracy of the system in the image plane, average transformation error introduced in (6.1) and sum of stabilization errors defined in (6.2) are used.

For verifying the accuracy of the system in three-dimensional space, the motion compensation error is defined. This error represents an Euclidean distance between the reference $\underline{v}^i_{k-n} \in \mathcal{V}_{k-n}$ and stabilized positions $\tilde{\underline{v}}^i_k \in \tilde{\mathcal{V}}_k$ of the evaluation points in $\mathbb{R}^3$. Averaged over the number of the

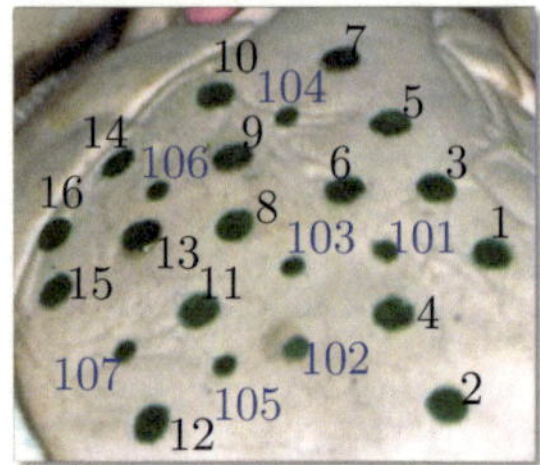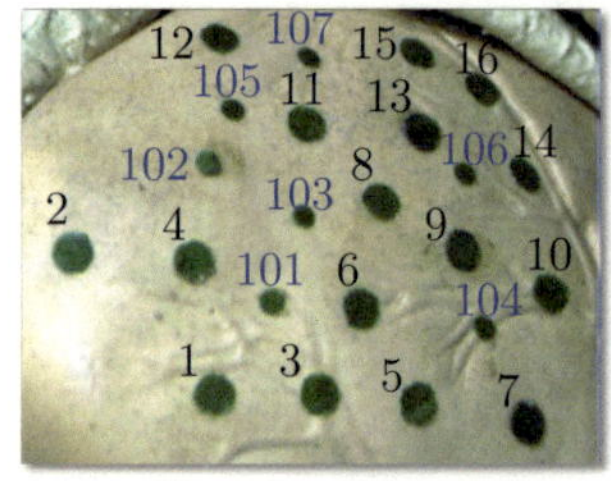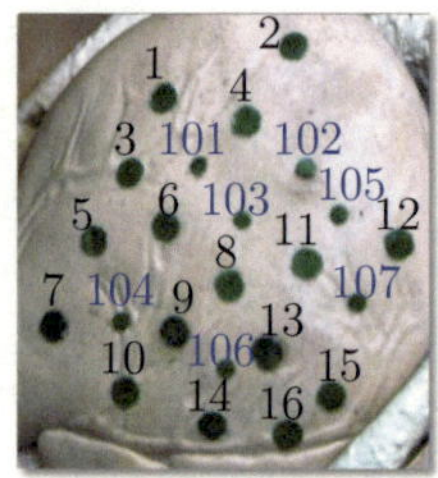

(a) Image of the first camera. (b) Image of the second camera. (c) Image of the third camera.

Figure 7.2: Images provided by a trinocular camera system. The markers used for the evaluation of the system are denoted by indices in blue. The markers, the measurements of which are processed by the system, are enumerated by indices in yellow.

evaluation points $N_{V_{k-n}}$ at every time step, it is computed according to

$$a_k^c = \sum_{i=1}^{N_{V_{k-n}}} \frac{\left\| \underline{v}_{k-n}^i - \underline{\tilde{v}}_k^i \right\|}{N_{V_{k-n}}}, \tag{7.1}$$

where the notation $\| \cdot \|$ indicates the Euclidean norm on $\mathbb{R}^3$. It should be noted that the reference positions of the evaluation points $\underline{v}_{k-n}^i \in V_{k-n}$ are determined by the surface reconstruction described in Section 5.4. The stabilized positions $\underline{\tilde{v}}_k^i \in \tilde{V}_k$ are obtained by the triangulation of those image features in the stabilized image that correspond to the evaluation points. For completeness of the description, the applied method of linear triangulation [80] is introduced in Appendix A.

7.2 Experimental Results

This section presents and discusses the experimental results of the system evaluation. It starts with the Section 7.2.1, where the three-dimensional reconstruction of the heart surface motion is testified. Then, Section 7.2.2 presents the experimental analysis of the accuracy and robustness of visual motion compensation.

7.2.1 Evaluation of Physics-Based Tracking

The aim of this section is to evaluate the capability of the system to predict and reconstruct the position of the object under observation in case of unexpected changes of object motion and loss of measurement information.

The sensitivity of the system to unexpected behavior of the object is evaluated in respect to the application in beating heart operations. For this purpose, the changes of cardiac rhythm, also called heart arrhythmia, are simulated in the experiments on the artificial heart. It should be mentioned that only the relative slow changes in the motion can be handled by the system. The system does not cope with extrasystoles leading to unexpected contractions of the heart, because the unpredictable and rash appearance of extrasystoles makes its simulation and handling very challenging.

For simulating heart arrhythmia, **two experiments** were carried out. For this purpose, the frequency of the pressure signal that determines the pressure inside of the heart is changed from 0.7 Hz to 1.2 Hz. Then, in order to verify the response of the system when the entire measurement information is lost, in the first experiment, total occlusions are simulated by neglecting all available measurements. In the second experiment, the heart surface is really occluded by the surgical instruments that leads to the loss of the measurement information.

1) Motion Reconstruction with Simulated Loss of Measurements   The results of estimating and predicting the heart motion at one of the evaluation points in case of simulated occlusions are illustrated in Fig. 7.3. The evaluation point under consideration is identified by the index 102 in Fig. 7.2. The gray shaded areas denote the time intervals where the entire measurement information is neglected by the system. In this case, the estimated position (in blue) of the evaluation point is equal to its predicted position (in red). The reason for this is that no measurements can be processed by the stochastic filter in the filter step, where the predicted state estimate should be corrected by incorporating the measurement information in the state estimation. Since the loss of the measurement information is simulated, the acquired measurements (in green) can be used for verification of the motion prediction.

As illustrated in Fig. 7.3, during the total occlusions, the predicted position of the evaluation point is similar to its measured position. This points out the consistency of the motion prediction with the measurement

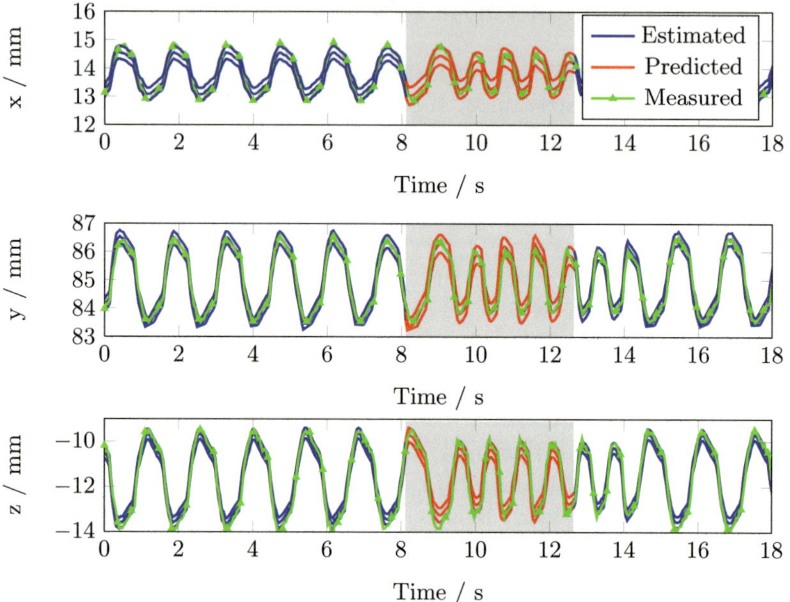

Figure 7.3: Motion reconstruction at the evaluation point during simulated heart arrhythmia. The predicted and estimated positions of the evaluation points are presented with 3σ bounds, which define that the true position of the evaluation point lies with $99,73\%$ probability inside these bounds. The areas shaded in gray denote the time intervals with simulated loss of the measurement information.

data. It should be noted that the predicted and estimated positions of the evaluation point are presented in Fig. 7.3 with 3σ bounds. These bounds define the region where the true position of the evaluation point can be found with $99,73\%$ probability.

2) Motion Reconstruction with Real Loss of Measurements Fig. 7.4 presents the results of the experiment with real occlusions. Here, the artificial heart also changes its cardiac rhythm. The shaded in gray regions denote the time intervals where the heart surface is partially occluded by different objects, such as surgical instruments.

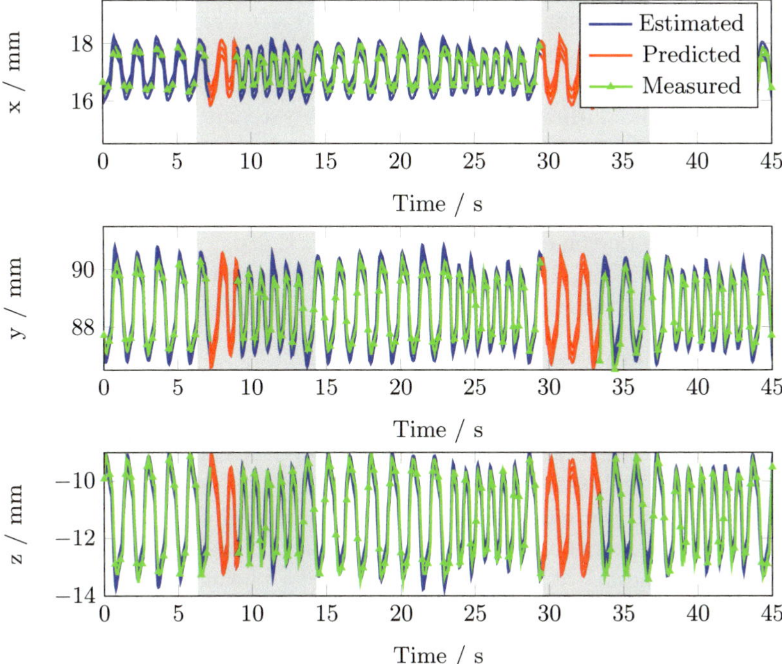

Figure 7.4: Motion reconstruction at the evaluation point by simulated changes of the cardiac rhythm. The time interval with partial loss of measurement information is denoted by the area shaded in gray. When the position of the point cannot be measured, its predicted position is still corrected by processing the remaining measurements. The predicted and estimated positions are presented with 3σ bounds giving $99,73\%$ probability that the true position of the point is inside these bounds.

As a result, only some of the image features corresponding to the landmarks could be measured. Therefore, only few measurement information is processed by the Gaussian filter in the filter step.

It is of advantage that when the position of the evaluation point cannot be measured, its predicted position is still improved by processing the available measurements. This occurs due to coupling between all points of the physical model, which is ensured by the physics-based approximation.

Discussion In summary, the experimental results reveal that there is no assumption on the periodic motion of the heart. The system reacts rapidly on the changing pressure. The main reason for that is in the consideration of the energy dissipation occurring in the real system due to, e.g., material damping. As discussed in Section 3.2, this effect is introduced in the model by the viscous material characteristics of the object and Rayleigh damping. The reaction of the system on the changes of pressure is rapid because of the assumption on the excitation of the heart that is primarily caused by the pressure inside of the heart chamber. This is appropriate for pressure regulated artificial heart.

Although certain research works state that heart excitation can be adequately represented by the pressure inside the cardiac chambers [191, 192], it is important to note that this assumption may lead to a slower reaction of the system on the changes of the real heart motion. Since the heart contraction and relaxation are caused by the muscle forces acting in a distributed fashion, the reaction time of the system will depend on the time delay between the changes of the pressure inside the heart chamber and the action of these forces. This problem can be tackled by the extending the system to the processing of the electrocardiogram signal, as introduced in the summary of Chapter 3.2.

A further advantage of the system is that it successfully bridges partial and even total occlusions. This is due to the fact that the system incorporates the a priori knowledge about the object's motion. This knowledge is continuously collected by the physical model and is essential for estimating the position of the heart by the loss of measurement information.

7.2.2 Evaluation of Visual Motion Compensation

This section illustrates the capability of the system to reconstruct and compensate the motion of the entire object under observation. Furthermore, the sensitivity of the motion compensation to the amount of the available measurement information is analyzed. Finally, the quantitative analysis of the accuracy of the system are given.

Functionalities of the System

In order to give an impression about the functionalities of the system, certain system results are presented in Fig. 7.5, Fig 7.6, and Fig. 7.7.

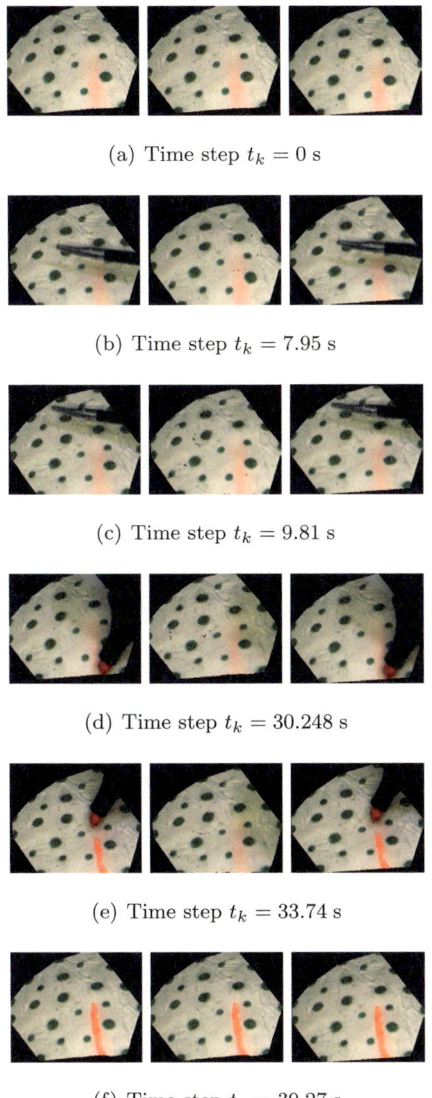

(a) Time step $t_k = 0$ s

(b) Time step $t_k = 7.95$ s

(c) Time step $t_k = 9.81$ s

(d) Time step $t_k = 30.248$ s

(e) Time step $t_k = 33.74$ s

(f) Time step $t_k = 39.27$ s

Figure 7.5: Current camera images (left), reconstructed (middle) and stabilized (right) images at selected time steps.

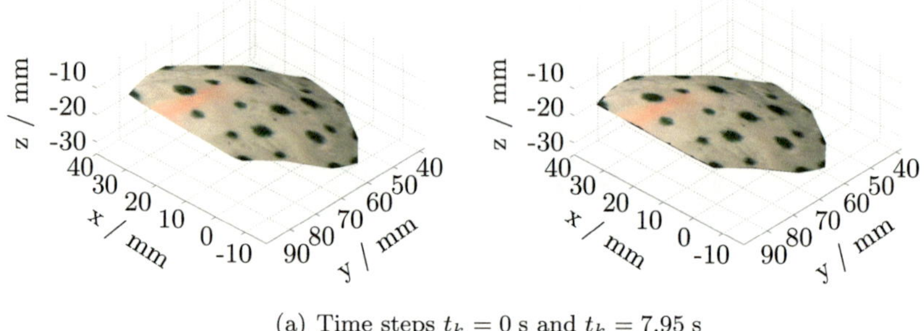

(a) Time steps $t_k = 0$ s and $t_k = 7.95$ s

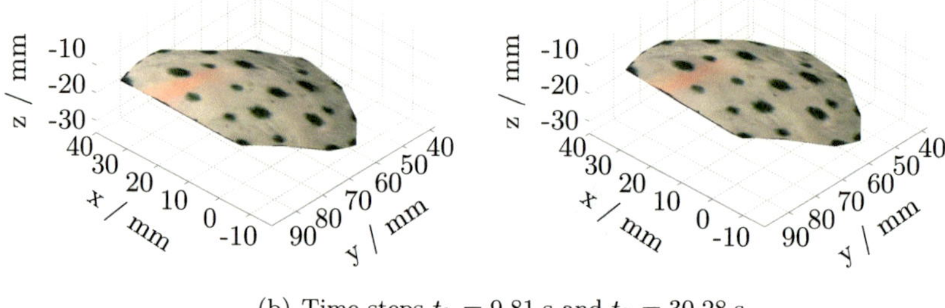

(b) Time steps $t_k = 9.81$ s and $t_k = 30.28$ s

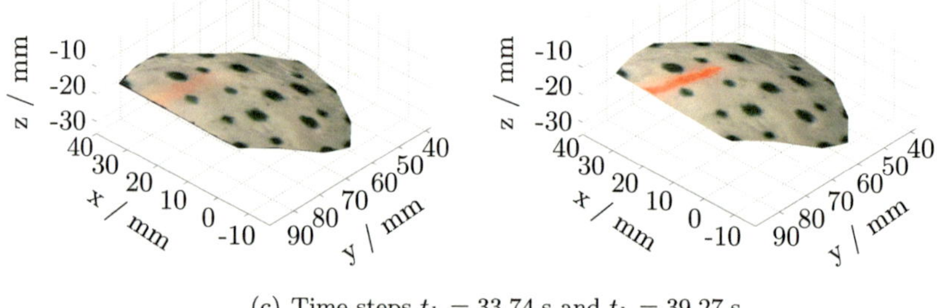

(c) Time steps $t_k = 33.74$ s and $t_k = 39.27$ s

Figure 7.6: Reconstructed surface of the artificial heart at selected time steps.

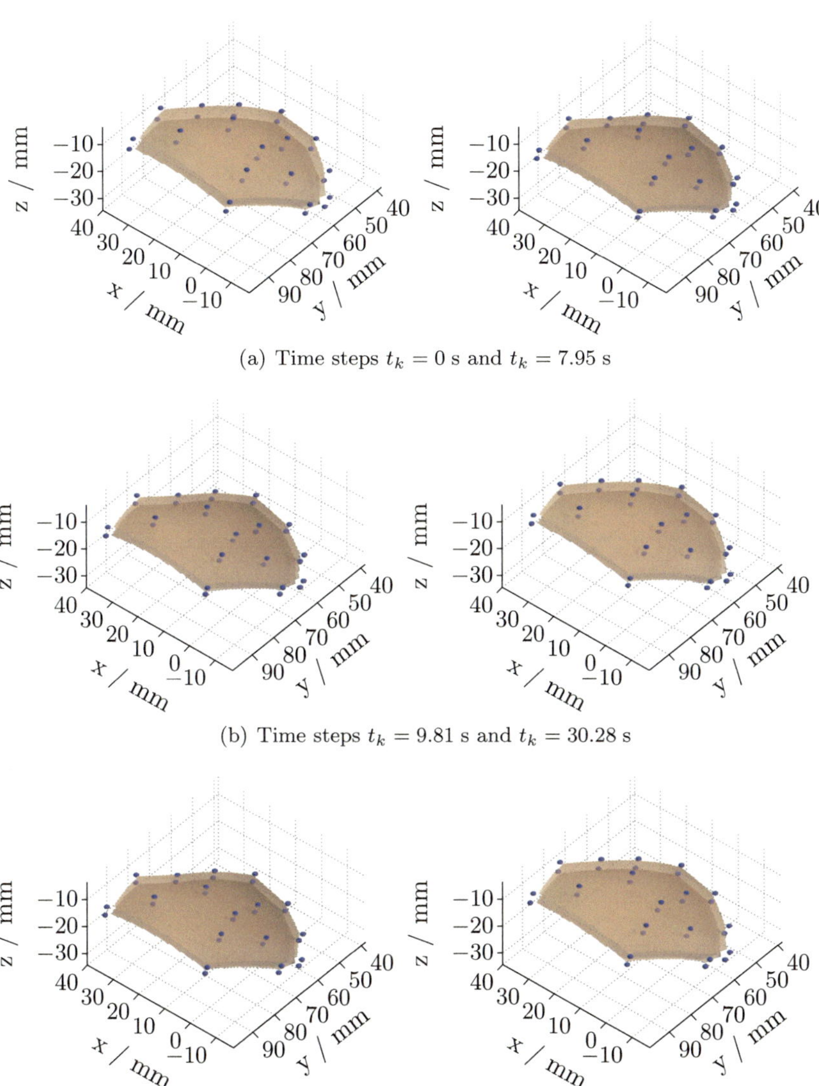

(a) Time steps $t_k = 0$ s and $t_k = 7.95$ s

(b) Time steps $t_k = 9.81$ s and $t_k = 30.28$ s

(c) Time steps $t_k = 33.74$ s and $t_k = 39.27$ s

Figure 7.7: Reconstructed heart wall at selected time steps. Two layers of the blue points represent the nodes of the discrete physical model.

The images presented in the fist column of Fig. 7.5 are selected from the image sequence provided by the first camera. In these images, the information about the appearance of the heart surface is partially lost due to the occlusion of the heart by other objects. The reconstruction of this information provided by the physics-based image transformation described in Section 5.4 is shown in the second column of Fig. 7.5. Furthermore, the stabilized images formed by the visual motion compensation introduced in Chapter 5 are illustrated in the third column of Fig. 7.5. These images present the current view of the moving heart surface as motionless. The objects occluding the heart surface, as well as changes of the heart color, remain visible in the stabilized image.

The position of the entire surface of the object estimated at every time step is depicted in Fig. 7.6, which clarifies the specialty of the system in reconstructing even the areas of the heart occluded by the other objects. For this purpose, the method introduced in Section 5.4 is applied.

Furthermore, as illustrated in Fig. 7.7, the volumetric physical model allows for reconstructing the current position of the entire heart wall, the surface of which is observed. Here, two layers of the blue points represent the nodes of the discrete physical model. All points between these nodes can be obtained by physics-based approximation according to (5.2) and (5.4). Therefore, in contrast to existing methods for motion compensation of the beating heart, this proposed system is able to reconstruct the motion of the object not only on its surface but also in the interior. This will significantly extend the surgeon's capabilities in manipulating the heart and navigating a surgical robot during beating heart operations.

Sensitivity to Loss of Measurements

The sensitivity of the system to the loss of the measurements is determined by a sequence of experiments, the results of which are gathered in Table 7.1 and Fig. 7.8, where the maximum of the average motion compensation is depicted in dependence on the number of available measurements.

The motion of the artificial heart is visually compensated by processing the same image sequences of the camera system, whereby the number of measurements extracted from these sequences is stepwise reduced. Hence, 16, 11, 7, 4, and finally, no measurements are processed by the stochastic filter. The accuracy of the motion compensation is determined at every time step by the motion compensation error (7.1) that is averaged over 7

Measurements	Motion compensation error in mm	
	Physics-based	Geometric
16	0.432	0.717
11	0.598	0.721
7	0.743	0.903
4	0.895	1.965
0	1.036	-

Table 7.1: Sensitivity of the physics-based and geometric motion compensation to the loss of the measurement information measured by the maximum of the motion compensation error that is averaged over 7 evaluation points at every time step. The maximum of this error is computed over the time period of 9 s.

evaluation points. For assessing the quality of the motion compensation over the time interval of 9 s, the maximum of this error is determined in the interval. This is motivated by the importance of high system accuracy at each time step.

The performance of the physics-based system is compared with the standard image warping method proposed in [8]. This method refers to the group of methods introduced in Section 1.3.1 that are based on the geometric image transformation. The current camera image is transformed to the stabilized image by means of a two-dimensional image transformation function that combines the affine transformation with the radial transformation. The latter transformation is determined by Gaussian radial basic functions with local support, including only the neighboring features of each extracted image feature.

To allow a fair and transparent comparison of the methods, the physics-based method for visual compensation is applied without the adaptation of the spatial resolution of the physical model by refinement of the discretization. Therefore, the standard method as well as the physics-based method perform on the same number of points.

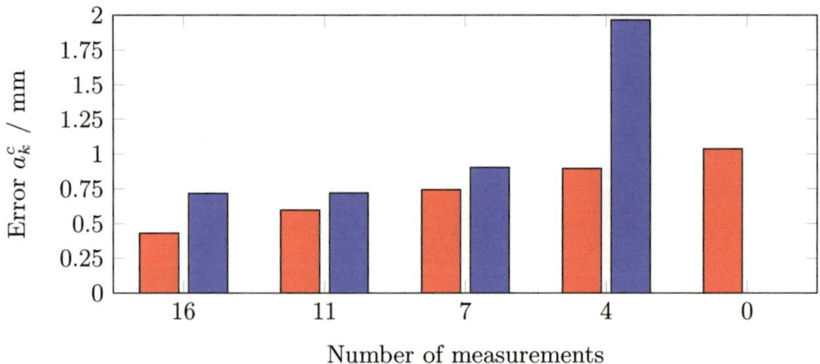

Figure 7.8: The physics-based motion compensation (red) has an advantage of lower dependence upon the amount of measurement information in comparison to the standard image warping method (blue). The levels of bars are determined by the maximum of average motion compensation error a_k^c in the time period of 9 s. The motion compensation error is averaged over the 7 evaluation points at every time step.

According to Table 7.1 and Fig. 7.8, the physics-based method has the advantage of lower dependence upon the loss of measurement information than the standard image warping method. The main reason for this is that more a priori information is incorporated in the physics-based method. This allows achieving a higher accuracy of the motion compensation even when less measurement information is available. Furthermore, this enables the reconstruction of the heart surface motion even in the case of complete loss of measurements that is not possible by the standard method.

Accuracy of Motion Compensation

The accuracy of the methods for motion compensation is testified by processing the data collected in three system runs. Each run consists of three image sequences, every of which contains 400 images provided by the trinocular camera system. The stabilized image sequences are obtained by processing these images in three ways: using standard image warping method [8] based on geometric image transformation, physics-based image transformation without and with feedback mechanism.

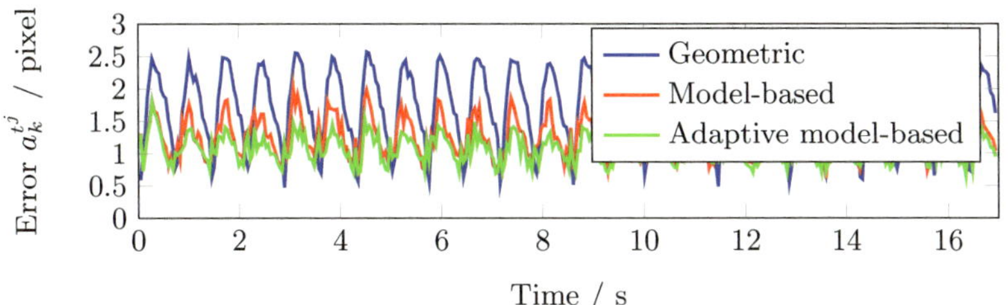

(a) Average transformation error at every time step

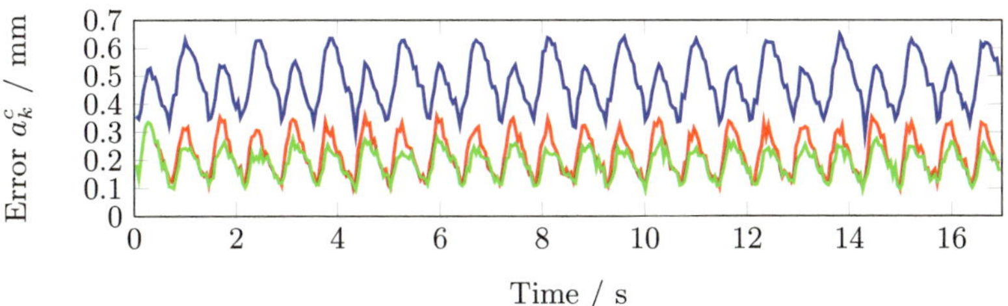

(b) Average motion compensation error at every time step.

Figure 7.9: Accuracy of geometric, physics-based, and adaptive physics-based methods measured by the transformation and motion compensation errors averaged over the 7 evaluation points at every time step. The adaptive physics-based method allows more accurate compensation of the large displacements of the object due to refined spatial resolution.

All methods are initialized on the same amount of points. However, the physics-based image transformation with feedback mechanism refines the spatial resolution of the motion compensation during the functionality of the system. For the purpose of the feedback, the stabilization error is computed over 23 frames, whereby the color values between 8 and 12 gray levels are binarized. This is why the adaptation of the physics-based motion compensation does not start until the first second. Commonly, there are 16 measurements available at every time step.

The accuracy of the considered methods is presented in Fig. 7.9, where the transformation error $a_k^{t^j}$ in the image sequence of the first camera of

Methods	Transformation error in pixel	
	Maximum	Mean
Geometric	2.568	1.734
Physics-based	2.031	1.270
Adaptive physics-based	1.576	1.036

Table 7.2: The accuracy of the stabilized image sequence is measured by the maximum and mean of the transformation error that is averaged over 7 evaluation points at every time step. Its maximum and mean are computed over the time period of 16 s.

the trinocular camera system as well as the motion compensation error a_k^c are depicted. These errors, computed according to (7.1) and (6.1) are averaged over 7 evaluation points and 3 runs at every time step.

Image Transformation Error Obviously, the adaptive physics-based motion compensation is of superior quality – according to Table 7.2, a maximum average transformation error in the considered time period of 16 s is 1.576 pixel. In comparison, the physics-based motion compensation without feedback mechanism provides the error of 2.031 pixel. The geometric approach achieves the error of 2.568 pixel.

Therefore, the maximum error of the adaptive physics-based motion compensation is 39% lower than the error of the geometric approach. The mean values of the average transformation error computed over the considered time period follow the similar trend as the maximum values of this error.

Motion Compensation Error Before comparing the accuracy of the different methods, it should be noted that the transformation error strongly depends on the position of the camera. If the heart moves primarily along the optical axis of the camera, all methods may provide comparable results. The reason for this is the loss of information about the motion of the object due to its projection on the image plane.

Fig. 7.9(b) illustrates the accuracy of the motion compensation in three-dimensional space. Here, the motion compensation error of the adaptive physics-based motion compensation averaged over 7 evaluation points is significantly smaller than that of the geometric approach. The maximum of this error in the considered time period is 0.281 mm, as introduced in Table 7.3.

Methods	Motion compensation error in mm	
	Maximum	Mean
Geometric	0.648	0.478
Physics-based	0.357	0.224
Adaptive physics-based	0.281	0.180

Table 7.3: The accuracy of the system is measured by the maximum and mean of the motion compensation error that is averaged over 7 evaluation points and is computed over the time period of 16 s.

The error is an impressive 57% lower than the error of the geometric approach, which is 0.648 mm. The main reasons for the high performance of the physics-based method are the incorporated physical knowledge and consideration of the model and measurement uncertainties. Furthermore, the 0.357 mm maximum error of the physics-based motion compensation without feedback mechanism is improved substantially by 12%, due to adaptation of the model's discretization and parameters.

It should be noted that errors are highest when the artificial heart is subjected to the largest displacements. The adaptive physics-based motion compensation allows for better compensation of these displacements due to higher spatial resolution of the physical model. According to Table 7.3, the mean values of the motion compensation error show a similar trend as the maximum values of this error.

Sum of Stabilization Errors When the transformation error and motion compensation error evaluate the quality of the system only on the selected evaluation points, the sum of stabilization errors $a_k^{s^{j,i}}$ given in Equation (6.2) provides an insight in the accuracy of the system also between these points.

Fig. 7.10 presents the sum of stabilization errors computed for one of the stabilized image sequences consisting of 400 images. With the aim of giving an impression about the changes that must be compensated, Fig. 7.10(a) depicts the sum of the stabilization errors between the reference image and the current image provided by a camera. The stabilization error of the geometric, physics-based, and physics-based motion compensation with feedback mechanism is determined by the color differences between the reference image and stabilized images. This error points out the remaining motion in the stabilized image sequence.

After the comparison between Fig. 7.10(b) and Fig. 7.10(c), it is evident that the stabilized image sequence provided by the physics-based visual motion compensation is of higher accuracy than the geometric approach. Furthermore, as depicted in Fig. 7.10(d), its accuracy is additionally improved by the feedback mechanism in the highlighted regions.

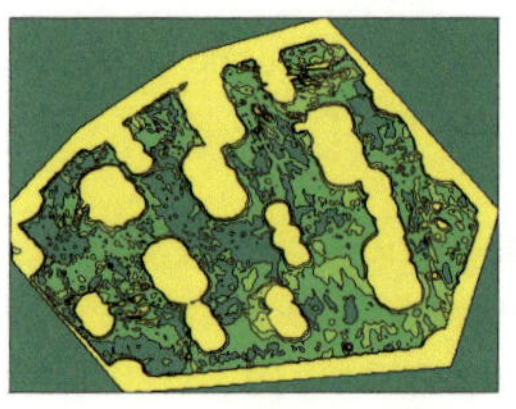

(a) Sum of color differences between camera images and the reference image.

(b) Geometric image transformation.

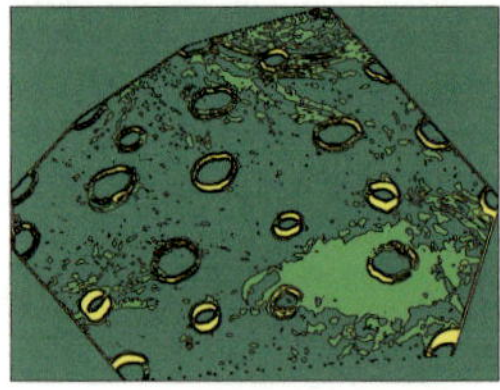

(c) Model-based image transformation without feedback.

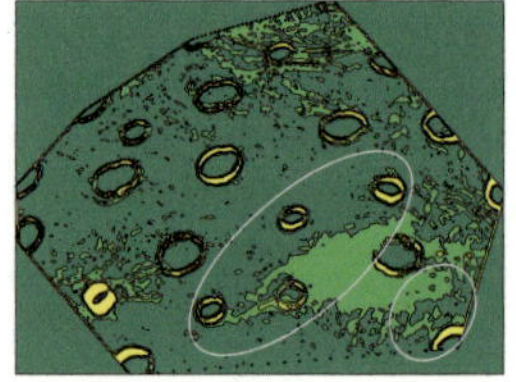

(d) Adaptive model-based image transformation with feedback mechanism.

32 63 94 125 156 187 218

(e) Levels of inaccuracies in green color of different intensities.

Figure 7.10: Sum of the stabilization error of the geometric, physics-based, and adaptive physics-based image stabilization in form of contour plots. This error is computed for one stabilized image sequence containing 400 images.

7.3 Summary

With respect to the computer-assisted beating heart surgery system, the performance of the proposed system for tracking and visual motion compensation is evaluated in experiments on the pressure regulated artificial beating heart, which simulates the motion of the mechanically stabilized real heart.

The functionality of the system is demonstrated in extensive experimental investigations testing the following system capabilities: (1) Handling of the relatively slow but unexpected changes of the object motion, (2) Reconstruction of the current camera image and formation of the stabilized image, as well as the estimation of the heart wall motion, (3) Sensitivity to the loss of measurement information, (4) Accuracy of motion compensation in the image plane and in three-dimensional space.

Contributions The physics-based motion compensation is compared to the motion compensation obtained by the standard method for image warping that describes the relationship between the current image and the reference image by the affine and radial transformation. As a result, the error of the adaptive physics-based motion compensation is 57% lower than the error of the geometric approach. There are two main reasons for the significantly better performance of the physics-base method: (1) Consideration of the measurement uncertainties and model imprecisions, (2) Incorporation of more a priori knowledge. The importance of the continuous monitoring of stabilization quality is emphasized by the fact that the error of the physics-based motion compensation is reduced by a significant 12% due to feedback mechanism. In the context of the beating heart surgery, the high accuracy of the system is of paramount importance for ensuring the safety of robotic operations.

As the acquisition of the camera measurements fails when the object is occluded, beating heart operations are unimaginable without loss of measurement information. For example, surgical instruments may often be in the field of the camera view. They manipulate the heart, which is observed by the camera. With this in mind, it is demonstrated that the proposed system is less sensitive to the loss of the measurement information than the standard method. Moreover, while the standard method fails when no measurements are available, the entire surface of the object as well as the current camera images can be reconstructed by the proposed system. The main reason for this consists in incorporating physical knowledge about the motion of the object under observation.

In addition, the unique property of the system to reconstruct the motion of the entire object by solely processing of camera data is presented. In this way, information about the motion in the interior of the heart wall, which is unaccessible to the methods commonly used for beating heart motion compensation, can be provided to a surgeon. This would significantly extend his capabilities in manipulating heart tissues and navigating a surgical robot. Furthermore, the reconstruction of the motion in the interior of the heart builds a foundation for modeling of surgical interventions.

Further Developments The next step towards the application of the proposed system in a computer-assisted beating heart surgery is its evaluation in in-vivo experiments. Here, one of the main challenges may be the extraction of measurement information, which is carried out by the proposed system in a simple way. Compared to the artificial heart, the heart surface is more textured and therefore, is more suitable for robust extraction of natural heart landmarks. However, this surface is of higher reflectance because it is smeared with blood, yielding a wet-like effect. This demands the preprocessing of the acquired images with the aim of compensating illuminations. Diverse approaches promise to tackle this problem [50, 118, 174]. Another issue that may cause difficulties is that the proposed system is object-specific due to its physics-based nature. Since the artificial beating heart is a poor imitation of the beating heart, it cannot be excluded that some corrections of the models, e.g., concerning the model excitation, are necessary for accurate estimation of the beating heart motion.

8 Conclusions and Future Work

In this research thesis, the physics-based framework for motion compensation of elastically deformable objects is developed. It can be applied in a wide range of applications of service and industrial robotics, as well as in video processing. In this thesis, the scope is concentrated mainly on the application to a beating heart robotic surgery system.

This system aims to facilitate a surgeon by operating on a beating heart. Apart from robotic navigation and manipulation, it performs the motion compensation of the beating heart. During an operation, it synchronizes the surgical instruments with the motion of the beating heart by exploiting the predicted position of the heart. At the same time, it represents the moving heart to a surgeon as motionless in order to improve the hand-eye coordination. This robotic system promises higher dexterity, lower error rate, and no fatigue. Furthermore, it makes long-distance teleoperations possible.

In order to enable the motion compensation, the accurate reconstruction and prediction of the position of the continuously changing entire elastic object is essential. This is complicated by the fact that the noisy camera measurements are available only at some discrete points. Furthermore, the cameras are very sensitive to occlusions and contaminations that lead to the loss of measurement information. The main challenge is the achievement of high accuracy substantial for safety relevant applications, with the efficiency necessary for the real-time operability of the system.

Contributions

The main theoretical contributions of the proposed methods for tracking and visual motion compensation include:

- Physical interpretation of the measurement data as a basis for both methods, which allows for the introduction of the physical knowledge in the reconstruction of the entire object deformation from the available measurements. This is of paramount importance for physically correct and highly accurate motion prediction as well as in case of highly inaccurate measurements or loss of measurement information.

- A sophisticated and computationally cheap novel physical model of the heart wall motion within the operation area, described by a distributed parameter system in the form of the system of stochastic partial differential equations. This raises the challenge of the mathematically complex formulation of boundary conditions.

- Balance between accuracy and computational complexity of the motion compensation by the combination of the simplified models of the physical object and the measurement system with the detailed analysis of the model imprecisions. The systematic and stochastic errors are systematically considered throughout the entire process of motion reconstruction.

- Mathematical derivation of all models involved in the system from the physical model, wherein the numerical solution of the system of stochastic partial differential equations is exploited. The physical model is spatially discretized by using the element-free method. This method represents the spatial domain of the model by a set of points without establishing the predefined connection between them. This significantly contributes to the efficiency of the system.

- Reconstruction of the entire elastic object by solely processing the camera data. Although the camera measurements about the motion of the object are available only at some surface points, even the interior of the deformable object can be reconstructed thanks to the volumetric model of the heart wall incorporated in the proposed methods. This allows for access to the motion of the object in areas that cannot be measured.

- Continuous monitoring of the quality of the system and its adaptive improvement by the feedback mechanism that incorporates the a posteriori estimation of the modeling errors. By these methods, the spatial resolution of all models involved in the system as well as the material properties of the model are refined in the areas where modeling inaccuracies are detected.

The performance of the system incorporating both methods is demonstrated by experimental studies on a pressure regulated artificial heart. The reconstruction of the entire deformable object as well as motion compensation has higher accuracy compared to the standard method based on the geometric image transformation. Furthermore, it copes success-fully with partial and even total occlusions. In contrast to the geometric approach, a lower number of measurements is required for achieving high motion compensation accuracy.

With respect to the application to the beating heart robotic surgery system, the system has the advantage of flexible adaptation to changing environmental conditions. The uncertainty bounds of the estimated heart surface position provide safety relevant information for control of the surgical robot. Thanks to the reconstruction of the motion of the entire object, the surgeon acquires information about the motion of the heart wall even in its interior. This will extend its capabilities in planning of intraoperative surgical interventions and navigation of the surgical robot.

Future Work

Since the enhancements of the individual components of the proposed framework have been discussed in the summary sections at the end of each chapter, long term research directions will be noted in this section. Further developments concern the application of the proposed methods in the beating heart robotic surgery system, as well as the generalization of the system for other applications.

Towards Computer-Assisted Beating Heart Surgery System

As regards the robotic surgery system, the proposed methods are intended to be evaluated in scenarios of real beating heart operations. Then, they are supposed to be integrated into the robotic surgery system aimed at navigation and control of surgical manipulators for performing surgical interventions.

Validation in Beating Heart Operations By concentrating on the reduction of the computational complexity of the motion compensation system, this work has not dealt with the sophisticated implementation of the proposed methods. Generally, the lower the frame rate of a data acquisition and processing is, the better the model for accurately reconstructing the heart motion must be. The reason for this is the insufficient amount of measurement data for compensating strong changes between acquired images. Therefore, the first steps towards the validation of the system in beating heart operations are an optimization of the real-time functionality of the system by efficient implementation of the algorithms, including their parallelization, as well as running of some calculations on the graphics processing unit. Some ideas with respect to the parallelization techniques can be found in [37, 94].

In the context of beating heart operations, the extension of the methods to handling surgical interventions is unavoidable. For example, the surgical instruments permanently interact with the heart wall by grasping the heart tissues, cutting, or suturating. The latter manipulations can be modeled by introducing the discontinuities in the heart model. To the best of our knowledge, although the discontinuous models are broadly investigated [3, 157], there are still no approaches for handling surgical intervention during beating heart surgery. The proposed volumetric model of the heart wall builds a good foundation for this, thanks to the fact that it approximates the motion of the heart wall also in the interior. The other challenge of modeling the surgical interventions is in the detection of the interactions between the heart tissues and surgical instruments. A basic approach to tackle this problem is to integrate the force sensors in surgical instruments, as proposed in [29, 40].

Another point that is identified as worthy for further investigations is the extension of the system for beating heart operations on a non-stabilized beating heart. This would significantly reduce the cost and time of beating heart operations due to lack of requirement for a mechanical stabilizer. The challenge here is that the non-stabilized heart has large deformations and more complex behavior. This demands that the physical model copes with these deformations. Currently, the model is restricted to small deformations and it is not rotation-invariant. This means that the large rotational deformations lead to the distortion of the model and to the unrealistic growth in its volume. The one way of solving this problem is to exploit Saint-Venant strain tensor [53] instead of Cauchy's strain tensor. However, this will lead to the high dimensional nonlinear system, which is not computationally feasible. The other way to guarantee the preservation of the volume by large deformations is corotational formulation [142, 143], which keeps the model linear and therefore, seems to be more suitable for this application.

Model-Predictive Control   Once the motion of the heart is predicted by the proposed method, the high quality control of the surgical manipulator becomes of high importance.

The model-predictive control methods [41] that, in contrast to classical methods, consider not only the current state of the surgical manipulator, but also its evaluation over the prediction horizon can meet these needs. These methods are especially advantageous due to the intrinsic compensation of the dead times [41] in real-time applications.

Furthermore, the constraints on the position of the manipulator defined by virtual features derived from the predicted position of the heart surface [238] can be included in the design of control in order to ensure the safety of the robotic operation. These features may also represent the software-generated forces influencing the motion of the robot, in order to improve the accuracy, the safety and the speed of the robot-assisted manipulations.

There are diverse concepts for incorporating the virtual features in the control of manipulator [1, 2, 31] that concern itself with guidance virtual fixtures and forbidden-region virtual fixtures. The first category may assist a surgeon in the navigation of the manipulator by moving the surgical instruments along desired path or heart surfaces. The forbidden-region virtual fixtures may prevent the manipulator from entering into critical regions of the heart.

Information Fusion and Sensor Management Considering that the a priori information leads to more accurate reconstruction of the heart wall, it is intended to enhance the proposed system with respect to a combination with additional information. One of the possibilities is to introduce the data from preoperative planning in the initialization of the physical model. The main challenge here, i.e., the registration of these data with the physical model, may be faced with methods [83, 224] for non-rigid body registration.

In addition, the combination with data provided by a medical imaging, like ultrasound, as well as by an electrocardiogram, or by other sensors in runtime, can further improve the quality of the motion compensation. For example, the information fusion with measurements provided by acceleration sensors [82, 144] would extend the capability of the system to capture the high frequent motion of the heart. The processing of the electrocardiogram signal would enable the prediction of the abnormal motion of the heart about 90 ms ahead, as stated in [45].

Furthermore, it is worth investigating the optimal locating of the artificial landmarks on the surface of the heart, to estimate the heart surface position as accurately as possible. Sensor management techniques [88,214] may deal with this problem. Such techniques aim at maximizing the informational content of the system by feasible configuration of the sensing modalities. The main challenge here is the consideration of the regions where the landmarks would disturb the surgical manipulations as constraints in sensor management.

Generalization for Other Applications

With respect to application of the proposed physics-based technique in video processing as well as in industrial or service robotics, the requirement on the available excitation of the physical model is disruptive. For compensating the motion of objects subjected to unknown excitation, the forces causing the deformation of the object under observation can be extracted from images, similarly to [141]. Furthermore, while in the above-mentioned applications multiple diverse objects can move in the scene under observation, the motion of every object should be compensated individually. Therefore, the accurate detection and classification of the objects in camera images is of high importance here.

A Linear Triangulation

The method of linear triangulation, e.g., introduced in [80] reconstructs the position of the point in $\mathbb{R}^3$ from its projections on the multiple camera images. Since the handling of two and more camera images by the linear triangulation is analogous, in this section, the measurement point is reconstructed by processing two camera images achieved by a camera system at the same time step.

Hence, the problem that the method of linear triangulation solves is finding the unknown position of the measurement point $\underline{l}^i = \left[x^{l^i}, y^{l^i}, z^{l^i} \right]^{\mathrm{T}} \in \mathbb{R}^3$ by processing these images. It is supposed that the positions of the image features $\underline{f}^i = \left[x^{f^i}, y^{f^i} \right]^{\mathrm{T}} \in \mathbb{R}^2$ and $\underline{f}^j = \left[x^{f^j}, y^{f^j} \right]^{\mathrm{T}} \in \mathbb{R}^2$ originated from the measurement point are known.

Therefore, the projection of the measurement point onto the image plane of the first and second cameras is described by equations

$$\underline{f}^i_k = \mathbf{P}^i \underline{l}^i \,, \quad \underline{f}^j_k = \mathbf{P}^j \underline{l}^i \,,$$

where the projection matrices $\mathbf{P}^i$ and $\mathbf{P}^j$ are determined by the cameras' calibration. The combination of the above equations, subject to constraints

$$\underline{f}^i \times \mathbf{P}^i \underline{l}^i = \underline{0} \,, \quad \underline{f}^j \times \mathbf{P}^j \underline{l}^i = \underline{0} \,,$$

results in the system of algebraic equations that is linear in $\underline{l}^i$ in terms of homogeneous coordinates

$$\mathbf{A}\underline{l}^i = \underline{0} \,. \tag{A.1}$$

The matrix $\mathbf{A}$ is defined by

$$\mathbf{A} = \begin{pmatrix} x^{f^i} \underline{p}^{3T}_i - \underline{p}^{1T}_i \\ y^{f^i} \underline{p}^{3T}_i - \underline{p}^{1T}_i \\ x^{f^j} \underline{p}^{3T}_j - \underline{p}^{1T}_j \\ y^{f^j} \underline{p}^{3T}_j - \underline{p}^{1T}_j \end{pmatrix} \,, \text{ with } \mathbf{P}^i = \begin{pmatrix} p_{11} & p_{12} & p_{13} & p_{14} \\ p_{21} & p_{22} & p_{23} & p_{24} \\ p_{31} & p_{32} & p_{33} & p_{34} \end{pmatrix} = \begin{bmatrix} \underline{p}^{1T}_i \\ \underline{p}^{2T}_i \\ \underline{p}^{3T}_i \end{bmatrix}$$

The system (A.1) is then solved by a singular values decomposition $\mathbf{A} = \mathbf{U}\mathbf{D}\mathbf{V}^{\mathrm{T}}$, where $\mathbf{U}$ and $\mathbf{V}$ are orthogonal matrices. The matrix $\mathbf{D}$ is diagonal with non-negative elements. The last column of $\mathbf{V}$, which is the eigenvector of $\mathbf{A}\mathbf{A}^{\mathrm{T}}$ corresponding to the smallest eigenvalue, represents the solution $\underline{l}^i$ of the system.

B Gaussian Filter

The aim of this section is to introduce the Gaussian filter [89] used in this thesis for simultaneous state and parameter estimation.

The Gaussian filter has been developed in [89]. Additionally, it is extended in [32] for state estimation of the conditionally linear models, whereby the state of the system is decomposed on linear and nonlinear substructures. In this section, the main equations of the Gaussian filter are introduced. It should be noted that this description is based on the above-mentioned works [32, 89]. The equations taken from these works are slightly modified for the system with additional noise.

It is assumed that the system model is a conditionally linear model of the form

$$\underline{y}_k = \underline{g}_k\left(\underline{x}_k^n\right) + \mathbf{G}_k\left(\underline{x}_k^n\right)\cdot\underline{x}_k^l + \underline{w}_k\,, \tag{B.1}$$

where the system state

$$\underline{x}_k = \begin{bmatrix} \underline{x}_k^l \\ \underline{x}_k^n \end{bmatrix}$$

is decomposed on linear $\underline{x}_k^l$ and nonlinear $\underline{x}_k^n$ substructures by using Rao-Blackwellization [63]. This decomposition allows to apply the Kalman filter equations for the computation of the linear substructure and Gaussian filter equations for the computation of the nonlinear substructure. It should be further noted that the vector function $\underline{g}_k$ and matrix function $\mathbf{G}_k$ depend on the nonlinear substructure. The stochastic perturbations collected in the vector $\underline{w}_k$ are additive zero-mean Gaussian.

The joint density of the vector $\underline{y}_k$ and of the system state $\underline{x}_k$ is equal to

$$f_k(\underline{y}_k,\underline{x}_k) = \delta\left(\underline{v}_k - \underline{g}_k(\underline{x}_k^n) - \mathbf{G}_k(\underline{x}_k^n) - \underline{w}_k\right) f_k(\underline{x}_k^l,\underline{x}_k^n)\,, \tag{B.2}$$

where the Dirac delta distribution is denoted by δ. The Gaussian filter approximates the joint density by a multivariate Gaussian density according to

$$f_k(\underline{y}_k,\underline{x}_k) \approx \mathcal{N}\left(\begin{bmatrix} \underline{\mu}_k^x \\ \underline{\mu}_k^y \end{bmatrix}, \begin{bmatrix} \mathbf{C}_k^{x,x} & \mathbf{C}^{x,y} \\ \mathbf{C}_k^{y,x} & \mathbf{C}^{y,y}, \end{bmatrix}\right)$$

where the first argument of the Gaussian function $\mathcal{N}$ denotes the mean of the density. The second argument expresses the covariance.

For computation of this density, the joint density of the linear and nonlinear substructures $f_k(\underline{x}_k^l, \underline{x}_k^n)$ involved in (B.2) can be separated according to Bayes law

$$f_k(\underline{x}_k^l,\ \underline{x}_k^n) = f_k(\underline{x}_k^n)f_k(\underline{x}_k^l \mid \underline{x}_k^n) , \tag{B.3}$$

where the conditional density is assumed to be Gaussian

$$f_k(\underline{x}_k^l \mid \underline{x}_k^n) \sim \mathcal{N}(\underline{\mu}_k(\underline{x}_k^n), \mathbf{C}_k^{l|n})$$

with mean and covariance

$$\underline{\mu}_k(\underline{x}_k^n) = \underline{\mu}_k^l + \mathbf{C}_k^{l,n} \cdot (\mathbf{C}_k^{n,n})^{-1} \cdot (\underline{x}_k^n - \underline{\mu}_k^n) ,$$
$$\mathbf{C}_k^{l|n} = \mathbf{C}_k^{l,l} - \mathbf{C}_k^{l,n} (\mathbf{C}_k^{n,n})^{-1} \cdot \mathbf{C}_k^{n,l} .$$

The Gaussian filter systematically approximates the density of the N dimensional nonlinear substructure by the Dirac-mixtures

$$f_k(\underline{x}_k^n) \approx \sum_{i=1}^{L} w_k \cdot \delta(\underline{x}_k^n - \underline{\mu}_{i,k}^n) , \tag{B.4}$$

with L Diracs weighted by $w_k = 1/L$. The total number of the sample points L is determined by N sets of D sample points placed along the N coordinate axes. For this approximation, the deterministic sampling schema presented in [89] is used. Consequently, the approximated density is computed by plugging (B.2), (B.3) and (B.4) in equation

$$f_k(\underline{y}_k) = \int_{\mathbb{R}} \int_{\mathbb{R}} f_k(\underline{y}_k, \underline{x}_k) \, \mathrm{d}\underline{x}_k^l \, \mathrm{d}\underline{x}_k^n .$$

This results in the representation of this density by the Gaussian mixture

$$f_k(\underline{y}_k) = \sum_{i=1}^{L} w_{k+1} \cdot \mathcal{N}(\underline{\mu}_{i,k}^y, \mathbf{C}_{i,k}^{y,y})$$

with mean and covariance

$$\underline{\mu}_{i,k}^y = \underline{g}_k(\underline{\mu}_{i,k}^n) + \mathbf{G}_k(\underline{\mu}_{i,k}^n) \cdot \underline{\mu}_k(\underline{\mu}_{i,k}^n) ,$$
$$\mathbf{C}_{i,k}^{y,y} = \mathbf{G}_k(\underline{\mu}_{i,k}^n) \cdot \mathbf{C}_k^{l|n} \cdot (\mathbf{G}_k(\underline{\mu}_{i,k}^n))^{\mathrm{T}}.$$

The first and second moments of the approximated density $f_k(\underline{y}_k)$ are determined by mean and covariance

$$\underline{\mu}_k^y = w_k \cdot \sum_{i=1}^{L} \underline{\mu}_{i,k}^y \,,$$

$$\mathbf{C}_k^{y,y} = \sum_{i=1}^{L} \left(w_k \cdot \mathbf{C}_{i,k}^{y,y} + w_k^s \cdot \left(\underline{\mu}_{i,k}^y - \underline{\mu}_k^y \right) \left(\underline{\mu}_{i,k}^y - \underline{\mu}_k^y \right)^{\mathrm{T}} \right) + \mathbf{C}_k^w \,, \tag{B.5}$$

where the weights $w_k^s = 1/D$ are used. Furthermore, the cross-covariance matrix is approximated by

$$\mathbf{C}_k^{x,y} = \sum_{i=1}^{L} \left(w_k \cdot \begin{bmatrix} \mathbf{0} \\ \mathbf{C}^{l \mid n} \, {}^{n}\mathbf{H}\left(\underline{\mu}_i^n\right)^{\mathrm{T}} \end{bmatrix} + w_k^s \cdot \left(\begin{bmatrix} \underline{\mu}_i^n \\ \underline{\mu}\left(\underline{\mu}_i^n\right) \end{bmatrix} - \underline{\mu}^x \right) \cdot \left(\underline{\mu}_i^y - \underline{\mu}_i^y \right)^{\mathrm{T}} \right). \tag{B.6}$$

It should be noted that the introduced equations can be applied in the prediction step as well as in the filter step of the Gaussian filter. A prerequisite for this is that the system model or measurement models can be converted in a conditionally linear form given in (B.1).

Lists of Figures, Tables, and Algorithms

List of Figures

List of Tables

List of Algorithms

Bibliography

[1] D. Aarno, S. Ekvall, and D. Kragíc. Adaptive Virtual Fixtures for Machine-Assisted Teleoperation Tasks. In *Proceedings of the 2005 IEEE International Conference on Robotics and Automation (ICRA 2005)*, volume 1, pages 897–903, Apr. 2005.

[2] J. J. Abbott, P. Marayong, and A. M. Okamura. Haptic Virtual Fixtures for Robot-Assisted Manipulation. In *Proceedings of the 12th International Symposium of Robotics Research (ISRR 2007)*, volume 28, pages 49–64, San Francisco, CA, USA, Oct. 2007.

[3] B. Adams and M. Wicke. Meshless Approximation Methods and Applications in Physics Based Modeling and Animation. *Eurographics 2009 Tutorial*, pages 1–27, Apr. 2009.

[4] Allied Vision Technologies GmbH. *PIKE Technical Manual V4.2.0*, Sept. 2009.

[5] K. H. An and M. J. Chung. 3D Head Tracking and Pose-Robust 2D Texture Map-Based Face Recognition using a Simple Ellipsoid Model. In *Proceedings of the 2008 IEEE/RSJ International Conference on Intelligent Robots and Systems (IROS 2008)*, pages 307–312, Nice, France, Sept. 2008.

[6] M. Anders and M. Hori. Stochastic Finite Element Method for Elasto-Plastic Body. *International Journal for Numerical Methods in Engineering*, 46(11):1897–1916, Dec. 1999.

[7] M. Anders and M. Hori. Three-Dimensional Stochastic Finite Element Method for Elasto-Plastic Bodies. *International Journal for Numerical Methods in Engineering*, 51(4):449–478, June 2001.

[8] N. Arad and D. Reisfeld. Image Warping Using few Anchor Points and Radial Funcions. *Computer Graphics Forum*, 14(1):35–46, Feb. 1994.

[9] G. M. Artmann and S. Chien. *Bioengineering in Cell and Tissue Research*. Springer, 2008.

[10] C. O. Arun, B. N. Rao, and S. M. Srinivasan. Stochastic Meshfree Method for Elasto-Plastic Damage Analysis. *Computer Methods in Applied Mechanics and Engineering*, pages 1–17, Apr. 2010.

[11] T. M. Atanackovic and A. Guran. *Theory of Elasticity for Scientists and Engineers*. Birkhäuser, 2000.

[12] S. N. Atluri. *The meshless Method (MLPG) for Domain and BIE Discretisation*. Tech Science Press, 2004.

[13] S. N. Atluri. *Methods of Computing Modelling in Engineering and the Science*, volume 1. Tech Science Press, 2005.

[14] S. N. Atluri, Z. D. Han, and H. T. Liu. Meshless Local Petrov-Galerkin (MLPG) Mixed Collocation Method for Elasticity Problems. *Computer Modeling in Engineering and Sciences (CMES)*, 14 (3):141–152, 2006.

[15] P. Azad, T. Gockel, and R. Dillmann. *Computer Vision–Das Praxisbuch*. Elektor-Verlag GmbH, 2007.

[16] I. Babuška, U. Banerjee, and J. E. Osborn. Survey of Meshless and Generalized Finite Element Methods: An Unified Approach. *Acta Numerica*, pages 1–125, July 2003.

[17] I. Babuška, F. Nobile, and R. Tempone. A Stochastic Collocation Method for Elliptic Partial Differential Equations with Random Input Data. *Journal on Numerical Analysis of Society for Industrial and Applied Mathematics (SIAM)*, 45(3):1005–1034, May 2007.

[18] I. Babuška, R. Tempone, and G. E. Zouraris. Galerkin Finite Element Approximations of Stochastic Elliptic Partial Differential Equations. *Journal on Numerical Analysis of Society for Industrial and Applied Mathematics (SIAM)*, 42(2):800–825, June 2005.

[19] T. Bader, A. Wiedemann, K. Roberts, and U. D. Hanebeck. Model-based Motion Estimation of Elastic Surfaces for Minimally Invasive Cardiac Surgery. In *Proceedings of the 2007 IEEE International Conference on Robotics and Automation (ICRA 2007)*, pages 2261–2266, Rome, Italy, Apr. 2007.

[20] Y. Bar-Shalom and T. E. Fortmann. *Tracking and Data Association*. Academic Press Professional, Inc, 1988.

[21] E. Bardinet, L. D. Cohen, and N. Ayache. Tracking and Motion Analysis of the Left Ventricle with Deformable Superquadrics. *Medical Image Analysis*, 1(2):129–149, Aug. 1996.

[22] A. Bartoli, E. von Tunzelmann, and A. Zisserman. Augmenting Images of Non-Rigid Scenes Using Point and Curve Correspondences. In *Proceedings of the 2004 IEEE International Conference on Computer Vision and Pattern Recognition (CVPR 2004)*, Washington, DC, USA, June 2004.

[23] O. Bebek and M. C. Çavuşoğlu. Intelligent Control Algorithms for Robotic-Assisted Beating Heart Surgery. *IEEE Transactions on Robotics*, 23(3):468–480, June 2007.

[24] O. Bebek and M. C. Çavuşoğlu. Whisker Sensor Design for Three Dimensional Position Measurement in Robotic Assisted Beating Heart Surgery. In *Proceedings of the 2007 IEEE International Conference on Robotics and Automation (ICRA 2007)*, pages 225–231, Roma, Apr. 2007.

[25] M. Becker, M. Ihmsen, and M. Teschner. Corotated SPH for Deformable Solids. In *Proceeding of the Eurographics Workshop on Natural Phenomena*, pages 27–34, Munich, Germany, Apr. 2009.

[26] M. Becker, H. Tessendorf, and M. Teschner. Direct Forcing for Lagrangian Rigid-Fluid Coupling. *IEEE Transactions on Visualization and Computer Graphics*, 15(3):493–503, June 2009.

[27] T. Belytschko, Y. Y. Lu, and L. Gu. Element-Free Galerkin Methods. *International Journal for Numerical Methods in Engineering*, 37:229–256, May 1994.

[28] S. Benhimane and E. Malis. Real-Time Image-Based Tracking of Planes Using Efficient Second-Order Minimization. In *Proceedings of the 2004 IEEE/RSJ International Conference on Intelligent Robot and Systems (IROS 2004)*, pages 943–948, Sendai, Japan, Oct. 2004.

[29] P. J. Berkelman, L. L. Whitcomb, R. H. Taylor, and P. Jensen. A Miniature Microsurgical Instrument Tip Force Sensor for Enhanced Force Feedback During Robot-Assisted Manipulation. *IEEE Ttansactions on Robotics and Automation*, 19(5):917–922, Oct. 2003.

[30] J. Bestel, F. Clément, and M. Sorine. A Biomechanical Model of Muscle Contraction. In *Proceeding of the 4th International Conference on Medical Image Computing and Computer-Assisted Intervention (MICCAI 2001)*, volume 2208 of *Lecture Notes in Computer Science*, pages 1159–1161, Utrecht, The Netherlands, Oct. 2001.

[31] A. Bettini, P. Marayong, S. Lang, A. M. Okamura, and G. D. Hager. Vision-Assisted Control for Manipulation using Virtual Fixtures. *IEEE Transactions Robotic*, 20:953–966, Nov. 2004.

[32] F. Beutler, M. F. Huber, and U. D. Hanebeck. Gaussian Filtering using State Decomposition Methods. In *Proceedings of the 12th International Conference on Information Fusion (Fusion 2009)*, Seattle, Washington, July 2009.

[33] D. Braess. *Finite Elemente: Theorie, Schnelle Löser und Anwendungen in der Elastizitätstheorie.* Springer, 2007.

[34] M. Bro-Nielsen. Finite Element Modeling in Surgery Simulation. In *Proceedings of the IEEE*, volume 86, pages 490–503, Mar. 1998.

[35] J. Brown, J. Rosen, M. Moreyra, M. Sinan, and B. Hannaford. Computer-Controlled Motorized Endoscopic Grasper for In Vivo Measurement of Soft Tissue Biomedical Characteristics. In J. D. Westwood, editor, *Medicine Meets Reality*. IOS Press, 2002.

[36] M. D. Buhmann. *Radial Basis Functions: Theory and Implementations.* Cambridge University Press, 2003.

[37] C. Burch. *Introduction to Parallel and Distributed Algorithms.* Hendrix College, 2009.

[38] J. C. Butcher. *Numerical Methods for Ordinary Differential Equations.* John Wiley and Sons, 2008.

[39] R. H. Byrd, M. E. Hribar, and J. Nocedal. An Interior Point Algorithm for Large-Scale Nonlinear Programming. *Journal on Optimization of Society for Industrial and Applied Mathematics (SIAM)*, 9(4):877–900, July 1999.

[40] D. J. Callaghan, M. M. McGrathy, and E. Coylez. Force Measurement Methods in Telerobotic Surgery: Implications for End-Effector

Manufacture. In *Proceedings of the 25th International Manufacturing Conference (IMC 2008)*, pages 389–398, Dublin, July 2008.

[41] E. F. Camacho and C. Bordons. *Model Predictive Control*. Springer, 2003.

[42] J. V. Candy. *Model-Based Signal Processing*. John Wiley and Sons, Inc., 2006.

[43] P. Cattin, H. Daveb, J. Grünenfelder, G. Szekelya, M. Turin, and G. Zünd. Trajectory of Coronary Motion and Its Significance in Robotic Motion Cancellation. *European Journal of Cardio-Thoracic Surgery*, 25:786–790, Feb. 2004.

[44] D. Causon and C. G. Mingham. *Introductory Finite Difference Methods for PDEs*. Ventus Publishing APs, 2010.

[45] M. C. Çavuşoğlu, J. Rotella, W. S. Newman, S. Choi, J. Ustin, and S. S. Sastry. Control Algorithms for Active Relative Motion Cancelling for Robotic Assisted Off-Pump Coronary Artery Bypass Graft Surgery. In *Proceedings of the 12th International Conference on Advanced Robotics (ICAR 2005)*, pages 431–436, Seattle, USA, July 2005.

[46] M. C. Çavuşoğlu, W. Williams, F. Tendick, and S. S. Sastry. Robotics for Telesurgery: Second Generation Berkeley/UCSF Laparoscopic Telesurgical Workstation and Looking towards the Future Applications. In *Proceedings of the 39th Allerton Conference on Communication, Control and Computing*, number 1, Monticello, IL, Oct. 2001.

[47] J. L. Chaboche, R. Girard, and A. Schaff. Numerical Analysis of Composite Systems by using Interphase/Interface Models. *Computational Mechanics*, 20(1-2):3–11, Jan. 1997.

[48] P.-G. Chassot, P. van der Linden, M. Zaugg, X. M. Mueller, and D. R. Spahn. Off-pump Coronary Artery Bypass Surgery: Physiology and Anaesthetic Management. *British Journal of Anaesthesia*, 92(3):400–413, 2004.

[49] C. W. Chen, T. S. Huang, and Y. C. Chen. Model-Based Estimation of Left Ventricle Motion. In *Proceedings of the 1991 International*

Conference on Acoustics, Speech, and Signal Processing (ICASSP 1991), volume 4, pages 2481–2484, Toronto, Ont., Canada, Apr. 1991.

[50] W. Chen, M. J. Er, and S. Wu. Illumination Compensation and Normalization for Robust Face Recognition using Discrete Cosine Transform in Logarithm Domain. IEEE Transactions on Systems, Man, and Cybernetics, 36(2):458–466, Apr. 2006.

[51] L. Chiari, U. D. Croce, A. Leardini, and A. Cappozzo. Human Movement Analysis Using Stereophotogrammetry. Part 2: Instrumental Errors. Gait and Posture, 21(2):197–211, Feb. 2005.

[52] P. Chomiac, S. Bose, and M. Page. Beating Heart Bypass Surgery. What Every Patient Should Know, 2001.

[53] P. G. Ciarlet. Mathematical Elasticity. Volume I: Three-Dimensional Elasticity, volume 20 of Studies in Mathematics and Its Applications. Elsevier Science Publishers B.V., 1988.

[54] S. Clavet, P. Beaudoin, and P. Poulin. Particle-Based Viscoelastic Fluid Simulation. In Proceeding of the 2005 Eurographics/ACM SIGGRAPH Symposium on Computer Animation (SCA 2005), pages 219–228, Los Angeles, California, USA, 2005.

[55] A. Corigliano. Formulation, Identification and Use of Interface Models in the Numerical Analysis of Composite Delamination. International Journal of Solids and Structures, 30(20):2779–2811, Mar. 1993.

[56] S. A. Cover, N. F. Ezquerra, J. F. O'Brien, R. Rowe, T. Gadacz, and E. Palm. Interactively Deformable Models for Surgery Simulation. IEEE Computer Graphics and Applications, 13(6):68–75, Nov. 1993.

[57] L. Cuvillon, J. Gangloff, M. de Mathelin, and A. Forgione. Toward Robotized Beating Heart TECABG: Assessment of the Heart Dynamics Using High-Speed Vision. In Proceedings of the 8th International Conference on Medical Image Computing and Computer-Assisted Intervention (MICCAI 2005), pages 551–558, Palm Springs, CA, Oct. 2005.

[58] M. H. Davis, A. Khotanzad, D. P. Flamig, and S. E. Harms. A Physics-Based Coordinate Transformation for 3-D Image Matching. *IEEE Transactions on Medical Imaging*, 16(3):317–328, June 1997.

[59] G. Deodatis. The Weighted Integral Method I: Stochastic Stiffness Matrix. *Journal of Engineering Mechanics*, 117(8):1851–1864, Aug. 1991.

[60] M. Desbrun and M.-P. Gascuel. Smoothed Particles: A New Paradigm for Animating Highly Deformable Bodies. In *Proceedings of the 1996 Eurographics Workshop on Computer Animation and Simulation (EGCAS 1996)*, pages 61–76, Poitiers, France, Aug. 1996.

[61] C. Detter, T. Deuse, F. Christ, D. H. Boehm, H. Reichenspurner, and B. Reichart. Comparison of Two Stabilizer Concepts for Off-Pump Coronary Artery Bypass Grafting. *The Annals of Thoracic Surgery*, 74:497–501, Apr. 2002.

[62] K. L. Dixon, A. Fung, and A. Celler. The Effect of Patient, Acquisition and Reconstruction Variables on Myocardial Wall Thickness as Measured from Myocardial Perfusion SPECT Studies. *IEEE Nuclear Science Symposium Conference Record*, 2004.

[63] A. Doucet, N. de Freitas, and N. Gordon. *Sequential Monte Carlo Methods in Practice*. Statistics for Engineering and Information Science. Springer, 2001.

[64] I. A. Essa and A. P. Pentland. Coding, Analysis, Interpretation, and Recognition of Facial Expressions. *IEEE Transactions on Pattern Analysis and Machine Intelligence*, 19(7):757–763, July 1997.

[65] R. Eymard, T. Gallouët, and R. Herbin. *Handbook of Numerical Analysis*, volume 7, chapter Finite Volume Methods, pages 713–1020. 2000.

[66] V. Falk. Manual Control and Tracking–A Human Factor Analysis Relevant for Beating Heart Surgery. *The Annals of Thoracic Surgery*, 74(2):624–628, Aug. 2002.

[67] G. E. Fasshauer. *Meshfree Approximation Methods with Matlab*, volume 6 of *Interdisciplinary Mathematical Sciences*. World Scientific Publishing Co. Pte. Ltd., 2007.

[68] G. S. Fishman. *Monte Carlo: Concepts, Algorithms, and Applications.* Springer Series in Operations Research. Springer, 1996.

[69] T. Fries and H. Matthies. Classification and Overview of Mesh-free Methods. Technical Report 2003-3, Technical University Braunschweig, July 2004.

[70] A. V. Gelder. Approximate Simulation of Elastic Membranes by Triangle Spring Meshes. *Journal of Graphics Tools*, 3(2):21–42, Feb. 1998.

[71] T. J. Gilhuly, S. E. Salcudean, and S. V. Lichtenstein. Evaluating Optical Stabilization of the Beating Heart. *IEEE Engineering in Medicine and Biology Magazine*, pages 133–140, Aug. 2003.

[72] R. A. Gingold and J. J. Monaghan. Smoothed Particle Hydrodynamics–Theory and Application to Non-Spherical Stars. *Monthly Notices of the Royal Astronomical Society*, 181:375–389, Nov. 1977.

[73] C. A. Glasbey and K. V. Mardia. A Review of Image Warping Methods. *Journal of Applied Statistics*, 25(2):155–171, 1998.

[74] F. Glover and M. Laguna. A Template of Scatter Search and Path Relinking. *Control and Cybernetics*, 29:653–684, 2000.

[75] M. S. Gockenbach. *Partial Differential Equations. Analytical and Numerical Methods.* Society for Industrial and Applied Mathematics, 2011.

[76] R. C. González and R. E. Woods. *Digital Image Processing.* Prentice Hall, 2008.

[77] M. Groeger, K. Arbter, and G. Hirzinger. *Medical Robotics*, chapter 10, pages 117–148. InTech, 2003.

[78] M. Gröger and G. Hirzinger. Image Stabilisation of the Beating Heart by Local Linear Interpolation. In K. R. Cleary and R. L. Galloway, editors, *Proceedings of SPIE Medical Imaging: Visualization, Image-Guided Procedures and Display*, volume 6141, pages 1–12, San Diego, USA, Feb. 2006.

[79] T. Harada, S. Koshizuka, and Y. Kawaguchi. Improvement in the Boundary Conditions of Smoothed Particle Hydrodynamics. *Computer Graphics and Geometry*, 9(3):2–15, 2007.

[80] R. Hartley and A. Zisserman. *Multiple View Geometry in Computer Vision*. Cambridge University Press, 2000.

[81] M. Hauth. *Visual Simulation of Deformable Models*. PhD thesis, Eberhards-Karls-Universität Tübingen, 2004.

[82] L. Hoff, O. Elle, M. Grimnes, S. Halvorsen, H. Alker, and E. Fosse. Measurements of Heart Motion Using Accelerometers. In *Proceedings of the 26th Annual International Conference of the IEEE Engineering in Medicine and Biology Society (EMBS 2004)*, San Francisco, CA, USA, Sept. 2004.

[83] M. Holden. A Review of Geometric Transformations for Nonrigid Body Registration. *IEEE Transactions on Medical Imaging*, 27(1): 111–128, Jan. 2008.

[84] R. D. Howe. Fixing the Beating Heart: Ultrasound Guidance for Robotic Intracardiac Surgery. In *Functional Imaging and Modeling of the Heart*, volume 5528/2009 of *Lecture Notes in Computer Science*, pages 97–103. Springer, 2009.

[85] J. Huang, X. Huang, and D. Metaxas. Optimization and Learning for Registration of Moving Dynamic Textures. In *Proceedings of the 11th IEEE International Conference on Computer Vision (ICCV 2007)*, pages 1–8, Rio de Janeiro, Oct. 2007.

[86] T. S. Huang and L. Tang. 3D Model-Based Video Coding: Computer Vision Meets Computer Graphics. In *Image Analysis Applications and Computer Graphics*, volume 1024/1995 of *Lecture Notes in Computer Science*, pages 1–8. Springer, 1995.

[87] X. Huang, N. A. Hill, J. Ren, G. Guiraudon, D. Boughner, and T. M. Peters. Dynamic 3D Ultrasound and MR Image Registration of the Beating Heart. In *Proceeding of the International Conference on Medical Image Computing and Computer-Assisted Intervention (MICCAI 2005)*, volume 3750/2005 of *Lecture Notes in Computer Science*, pages 171–178, Palm Springs, CA, USA, Oct. 2005.

[88] M. F. Huber. *Probabilistic Framework for Sensor Management*. PhD thesis, Universität Karlsruhe (TH), 2009.

[89] M. F. Huber and U. D. Hanebeck. Gaussian Filter based on Deterministic Sampling for High Quality Nonlinear Estimation. In *Proceedings of the 17th IFAC World Congress (IFAC 2008)*, volume 17, Seoul, Korea, July 2008.

[90] J. R. Huddle and D. A. Wismer. Degradation of Linear Filter Performance due to Modeling Error. *IEEE Transactions Automatic Control*, 13(4):421–423, Aug. 1968.

[91] T. J. R. Hughes. *The Finite Element Method. Linear Static and Dynamic Finite Element Analysis*. Dover Publications, 2000.

[92] S. A. Hutchinson, G. D. Hager, and P. I. Corke. A Tutorial on Visual Servo Control. *IEEE Transactions on Robotics and Automation*, 12 (5):651–670, Oct. 1996.

[93] P. A. Iaizzo, editor. *Handbook of Cardiac Anatomy, Physiology, and Devices (Current Clinical Oncology)* . Humana Press, 2005.

[94] N. Idrees and S. S. G. Pfister. Parallelization of Modular Algorithms. *Journal of Symbolic Computation*, 46:672–684, Jan. 2011.

[95] S. Jacobs, D. Holzhey, B. B. Kiaii, J. F. Onnasch, T. Walther, F. W. Mohr, and V. Falk. Limitations for Manual and Telemanipulator-Assisted Motion Tracking– Implications for Endoscopic Beating-Heart Surgery. *The Annals of Thoracic Surgery*, 76:2029–2036, May 2003.

[96] D. L. James and D. K. Pai. ArtDefo. Accurate Real Time Deformable Objects. In *Proceedings of the 26th Annual Conference on Computer Graphics and Interactive Techniques (SIGGRAPH 1999)*, pages 65–72, Los Angeles, CA, USA, Aug. 1999.

[97] K. O. Johnson, R. K. Robison, and J. G. Pipe. Rigid Body Motion Compensation for Spiral Projection Imaging. *IEEE Transactions on Medical Imaging*, 30(3):655–665, Mar. 2011.

[98] E. Jouini, J. Cvitanić, and M. Musiela, editors. *Option Pricing, Interest Rates and Risk Management*. Handbooks in Mathematical Finance. Cambridge University Press, 2001.

[99] S. J. Julier, J. K. Uhlmann, and H. F. Durrant-Whyte. A New Approach for Filtering Nonlinear Systems. In *Proceedings of the 1995 American Control Conference (ACC 1995)*, volume 3, pages 1628–1632, Seattle, WA , USA, June 1995.

[100] F. Jurie and M. Dhome. Hyperplane Approximation for Template Matching. *IEEE Transactions on Pattern Analysis and Machine Intelligence*, 24(7):996–1000, July 2002.

[101] R. E. Kalman. A New Approach to Linear Filtering and Prediction Problems. *Transactions of the American Society of Mechanical Engineers (ASME)–Journal of Basic Engineering*, 82:35–45, Feb. 1960.

[102] E. J. Kansa. Multiquadrics–A Scattered Data Approximation Scheme with Applications to Computational Fluid Dynamics–II Solution to Parabolic, Hyperbolic and Elliptical Partial Differential Equations. *Computer and Mathematics with Applications*, 19(8-9): 127–145, 1990.

[103] G. E. Karniadakis and S. J. Sherwin. *Numerical Mathematics and Scientific Computations. Spectral / Hp Element Methods for Computational Fluid Dynamics*. Oxford University Press, 2005.

[104] J. H. Kaspersen, T. Langø, and F. Lindseth. Computer-Aided Interventions. In K. Zieliński, M. Duplaga, D. Ingram, K. J. Hannah, and M. J. Ball, editors, *Information Technology Solutions for Healthcare*, Health Informatics, pages 271–287. Springer, 2006.

[105] M. Kass, A. Witkin, and D. Terzopoulos. Snakes: Active Contuour Models. *International Journal of Computer Vision*, 1(4):321–331, 1988.

[106] M. Kauer, V. Vuskovic, J. Dual, G. Szekely, and M. Bajka. Inverse Finite Element Characterization of Soft Tissues. In *Proceedings of the 4th International Conference on Medical Image Computing and Computer-Assisted Intervention (MICCAI 2001)*, volume 2208/2001 of *Lecture Notes in Computer Science*, pages 128–136, Utrecht, The Netherlands, Oct. 2001.

[107] A. D. Kiureghian and P. Liu. Structural Reliability Under Incomplete Probability Information. *Journal of Engineering Mechanics*, 112(1):85–104, Jan. 1986.

[108] R. E. Klabunde. *Cardiovascular Physiology Concept.* Lippincott Williams and Wilkins, 2011.

[109] R. Klette, K. Schlüns, and A. Koschan. *Computer Vision. Three-Dimensional Data from Images.* Springer, 1998.

[110] D. Koks and S. Challa. An Introduction to Bayesian and Dempster-Shafer Data Fusion. Dsto-tr-1436, Australian Government. Department of Defence, Nov. 2005.

[111] D. Kumar and C. Jawahar. Visual Servoing in Presence of Non-Rigid Motion. In *Proceedings of the 18th IEEE International Conference on Pattern Recognition (ICPR 2006)*, pages 655–658, Hong Kong, Sept. 2006.

[112] C. Kurz, T. Thormählen, and H. Seidel. Scene-Aware Video Stabilization by Visual Fixation. In *Proceedings of the 6th European Conference on Visual Media Production (CVMP 2009)*, pages 1–6, London, UK, Nov. 2009.

[113] A. Kurzhanski and I. Vályi. *Ellipsoidal Calculus for Estimation and Control.* Birkhäuser, 1997.

[114] W. W. Lau, N. A. Ramey, J. J. Corso, N. V. Thakor, and G. D. Hager. Stereo-Based Endoscopic Tracking of Cardiac Surface Deformation. In *Proceedings of the 7th International Conference on Medical Image Computing and Computer-Assisted Intervention (MICCAI 2004)*, pages 494–501, Rennes, France, Sept. 2004.

[115] L. P. Lebedev and M. J. Cloud. *Introduction to Mathematical Elasticity.* World Scientific Pub. Co., 2009.

[116] J. Lee, A. Eskandarian, and Y. Chen. *Meshless Methods in Solid Mechanics.* Springer, 2006.

[117] S. Lee, G. Wolberg, and S. Y. Shin. Scattered Data Interpolation with Multilevel B-splines. *IEEE Transactions on Visualization and Computer Graphics*, 3(3):228–244, Sept. 1997.

[118] S. W. Lee and S. W. Lee. SVDD-Based Illumination Compensation for Face Recognition. *Advances in Biometrics*, 2:323–337, 2008.

[119] Y. Lee, D. Terzopoulos, and K. Waters. Constructing Physics-Based Facial Models of Individuals. In *Proceedings of the Graphics Interface 1993 Conference*, pages 1–8, Toronto, ON, Canada, 1993.

[120] Y. Lee, D. Terzopoulos, and K. Waters. Realistic Modeling for Facial Animation. In *Proceedings of the 22nd Annual Conference on Computer Graphics and Interactive Techniques (SIGGRAPH 1995)*, pages 55–62, Los Angeles, CA, USA, Aug. 1995.

[121] V. A. Leitao and C. T. Fernandes. On a Multi-Region Trefftz Collocation Method for Plate Bending. In *Proceedings of the 4th World Congress on Computational Mechanics*, pages 1–17, Buenos Aires, Argentina, 1998.

[122] M. Lemma, A. Mangini, A. Redaelli, and F. Acocella. Do Cardiac Stabilizers Really Stabilize? Experimental Quantitative Analysis of Mechanical Stabilization. *Interactive Cardio Vascular and Thoracic Surgery*, 4:222–226, Mar. 2005.

[123] A. Lempel and J. Ziv. On the Complexity of Finite Secuences. *IEEE Transactions on Information Theory*, 22(1), 1976.

[124] S. Li and W. K. Liu. *Meshfree Particle Methods.* Springer, 2007.

[125] M. E. Liggins, D. L. Hall, and J. Llinas. *Handbook of Multisensor Data Fusion. Theory and Practice.* CRC Press, 2009.

[126] G. R. Liu. *Mesh Free Methods: Moving Beyond the Finite Element Methods.* CRC Press, 2003.

[127] G. R. Liu. *An Introduction to Meshfree Methods and Their Programming.* Springer, 2005.

[128] G. R. Liu and M. B. Liu. *Smoothed Particle Hydrodynamics. A Meshfree Particle Method.* World Scientific Pub. Co., 2003.

[129] P. Liu, H. Wang, X. Wu, and W. Qiao. Motion Compensation Based Detecting and Tracking Targets in Dynamic Scene. In *Proceedings of the 2010 IEEE International Conference on Measuring Technology and Mechatronics Automation (ICMTMA 2010)*, pages 703–706, Changsha City, Mar. 2010.

[130] W. Liu, T. Belytschko, and A. Mani. Probabilistic Finite Elements for Nonlinear Structural Dynamics. *Computer Methods in Applied Mechanics and Engineering*, 56(1):61–86, July 1986.

[131] W. Liu, T. Belytschko, and A. Mani. Random Field Finite Elements. *International Journal for Numerical Methods in Engineering*, 23(10): 1831–1845, June 1986.

[132] W. Liu and E. Ribeiro. A Survey on Image-Based Continuum-Body Motion Estimation. *Image and Vision Computing*, 29:509–523, 2011.

[133] B. D. Lucas and T. Kanade. An Iterative Image Registration Technique with an Application to Stereo Vision. In *Proceedings of the 7th International Joint Conference on Artificial Intelligence (IJCAI 1981)*, volume 2, pages 674–679, San Francisco, CA, USA, 1981.

[134] L. B. Lucy. A Numerical Approach to the Testing of the Fission Hypothesis. *Astronomical Journal*, 82:1013–1024, Dec. 1977.

[135] H. Maaß and U. Kühnapfel. Noninvasive Measurement of Elastic Properties of Living Tissue. In *Proceedings of the 13th International Congress on Computer Assisted Radiology (CARS 1999)*, pages 865–870, Paris, France, June 1999.

[136] P. Mahalanobis. On the Generalised Distance in Statistics. In *Proceedings of the National Institute of Science*, volume 2, pages 49–55, 1936.

[137] R. P. S. Mahler. *Statistical Mutlisource-Multitarget Information Fusion*. Artech House, Inc., 2007.

[138] G. Matthies, C. Brenner, C. Bucher, and C. Soares. Uncertainties in Probabilistic Numerical Analysis of Structures and Solids–Stochastic Finite Elements. *Structural Safety*, 19(3):283–336, May 1997.

[139] T. McInerney and D. Terzopoulos. A Dynamic Finite Elemente Surface Model for Segmentation and Tracking in Multidimensional Medical Images with Application to Cardiac 4D Image Analysis. *Computerized Medical Imaging and Graphics*, 19(1):69–83, May 1995.

[140] T. McInerney and D. Terzopoulos. Deformable Models in Medical Image Analysis: A Survey. *Medical Image Analysis*, 1(2):91–108, 1996.

[141] D. N. Metaxas. *Physics Based Deformable Models*. Springer, 1997.

[142] J. Mezger, B. Thomaszewski, S. Pabst, and W. Straßer. Interactive Physically-Based Shape Editing. In *Proceedings of the 2008 ACM Symposium on Solid and Physical Modeling (SPM 2008)*, pages 79–89, 2008.

[143] M. Müller, R. Keiser, A. Nealen, M. Pauly, M. Gross, and M. Alexa. Point Based Animation of Elastic, Plastic and Melting Objects. In *Proceedings of the 2004 ACM SIGGRAPH / Eurographics Symposium on Computer Animation (SCA 2004)*, pages 141–151, Grenoble, France, Aug. 2004.

[144] S. B. Moberg and J. D. Causey. Implantable Leads Incorporating Cardiac Wall Acceleration Sensors and Method of Fabrication. Patent A61B 502, 1997.

[145] J. J. Monaghan. An Introduction to SPH. *Computer Physics Communications*, 48(1):89–96, 1988.

[146] J. J. Monaghan. Smoothed Particle Hydrodynamics. *Annual Review of Astronomy and Astrophysics*, 30:543–574, 1992.

[147] D. R. Morrell and W. C. Stirling. Set-Valued Filtering and Smoothing. *IEEE Transactions on Systems, Man and Cybernetics*, 21(1): 184–193, Feb. 1991.

[148] M. Mueller and M. Gross. Interactive Virtual Materials. In *Proceedings of 2004 Graphics Interface (GI 2004)*, pages 239–246, Waterloo, Ontario, Canada, 2004. Canadian Human-Computer Communications Society.

[149] H. Nakagaki, K. Kitagaki, T. Ogasawara, and H. Tsukune. Study of Deformation and Insertion Tasks of a Flexible Wire. In *Proceedings of the 1997 IEEE International Conference on Robotics and Automation (ICRA 1997)*, volume 3, pages 2397–2402, Albuquerque, NM, USA, Apr. 1997.

[150] Y. Nakamura, K. Kishi, and H. Kawakami. Heartbeat Synchronization for Robotic Cardiac Surgery. In *Proceedings of the 2001 IEEE International Conference on Robotics and Automation (ICRA 2001)*, pages 2014–2019, Seoul, Korea, May 2001.

[151] B. Nayroles, G. Touzot, and P. Villon. Generalizing the Finite Element Method: Diffuse Approximation and Diffuse Elements. *Computational Mechanics*, 10:307–318, 1992.

[152] V. P. Nguyen, T. Rabczuk, S. Bordas, and M. Duflot. Meshless Methods: A Review and Computer Implementation Aspects. *Mathematics and Computers in Simulation*, 79(3):763–813, Dec. 2008.

[153] H. Niederreiter. *Random Number Generation and Quasi-Monte Carlo Methods*, volume 63 of *CBMS-NSF Regional Conference Series in Applied Mathematics*. Society for Industrial and Applied Mathematics (SIAM), 1992.

[154] B. Noack, V. Klumpp, D. Brunn, and U. D. Hanebeck. Nonlinear Bayesian Estimation with Convex Sets of Probability Densities. In *Proceedings of the 11th International Conference on Information Fusion (Fusion 2008)*, pages 1–8, Cologne, Germany, July 2008.

[155] B. Noack, V. Klumpp, and U. D. Hanebeck. State Estimation with Sets of Densities considering Stochastic and Systematic Errors. In *Proceedings of the 12th International Conference on Information Fusion (Fusion 2009)*, Seattle, Washington, July 2009.

[156] A. Noce, J. Triboulet, and P. Poignet. Efficient Tracking of the Heart Using Texture. In *Proceedings of the Annual International Conference of the IEEE Engineering in Medicine and Biology Society (EMBS 2007)*, pages 4480–4483, Lyon, France, Aug. 2007.

[157] J. Oliver, A. E. Huespe, and P. J. Sánchez. A Comparative Study on Finite Elements for Capturing Strong Discontinuities: E-FEM vs X-FEM. *Computational Methods in Applied Mechanical Engineering*, 195:4732–4752, Sept. 2006.

[158] S. R. Ommen, R. A. Nishimura, C. P. Appleton, F. A. Miller, J. K. Oh, M. M. Redfield, and A. J. Tajik. Clinical Utility of

Doppler Echocardiography and Tissue Doppler Imaging in the Estimation of Left Ventricular Filling Pressures. A Comparative Simultaneous Doppler-Catheterization Study. *Circulation. Journal of the American Heart Assotiation*, 102:1788–1794, 2000.

[159] T. Ortmaier. *Motion Compensation in Minimally Invasive Robotic Surgery.* PhD thesis, Lehrstuhl für Realzeit-Computersysteme, Technische Universität München, 2002.

[160] T. Ortmaier, M. Gröger, D. H. Boehm, V. Falk, and G. Hirzinger. Motion Estimation in Beating Heart Surgery. *IEEE Transactions on Biomedical Engineering*, 52(10):1729–1740, Oct. 2005.

[161] J. Park, D. Metaxas, and L. Axel. Volumetric Deformable Models with Parameter Functions: A New Approach to the 3D Motion Analysis of the LV from MRI-SPAMM. In *Proceedings of the 5th International Conference on Computer Vision (ICCV 1995)*, pages 700–705, Cambridge, MA , USA, June 1995.

[162] F. I. Parke. Parameterized Models for Facial Animation. *IEEE Computer Graphics and Applications*, 2(9):61–68, Nov. 1982.

[163] A. M. Pertsov, J. M. Davidenko, R. Salomonsz, W. T. Baxter, and J. Jalife. Spiral Waves of Excitation Underlie Reentrant Activity in Isolated Cardiac Muscle. *Circulation Research*, 72:631–650, 1993.

[164] M. Pilu. A Direct Method for Stereo Correspondence Based on Singular Value Decomposition. In *Proceedings of the 1997 IEEE Compuer Society Conference on Computer Vision and Pattern Recognition (CVPR 1997)*, pages 261–266, San Juan , Puerto Rico, June 1997.

[165] S. M. Platt and N. I. Badler. Animating Facial Expressions. In *Proceedings of the 8th Annual Conference on Computer Graphics and Interactive Techniques (SIGGRAPH 1981)*, volume 15, pages 245–252, Aug. 1981.

[166] P. Pratt, M. V. Scarzanella, D. Stoyanov, and G.-Z. Yang. Dynamic Guidance for Robotic Surgery using Image-Constrained Biomechanical Models. In *Proccedings of the 13th International Conference on Medical Image Computing and Computer Assisted Interventions (MICCAI 2010)*, volume 1, pages 77–85, Beijing, China, Sept. 2010.

[167] Z. Q. Qu. *Model Order Reduction Techniques: with Applications in Finite Element Analysis.* Springer, 2004.

[168] S. Rahman and B. Rao. An Elementfree Galerkin Method for Probabilistic Mechanics and Reliability. *International Journal of Solids and Structures*, 38:9313–9330, Aug. 2001.

[169] S. Rahman and B. N. Rao. A Perturbation Method for Stochastic Meshless Analysis in Elastostatics. *International Journal for Numerical Methods in Engineering*, 50:1969–1991, June 2001.

[170] N. A. Ramey, J. J. Corso, W. W. Lau, D. Burschka, and G. D. Hager. Real-Time 3D Surface Tracking and Its Applications. In *Proceedings of the 2004 IEEE Compuer Society Conference on Computer Vision and Pattern Recognition Workshop (CVPRW 2004)*, volume 3, pages 1–8, June 2004.

[171] J. W. S. Rayleigh. *Theory of Sound*, volume 1. MacMillan, 1894.

[172] Q. Ren, S. Nishioka, H. Shirato, and R. I. Berbeco. Adaptive Prediction of Respiratory Motion for Motion Compensation Radiotherapy. *Physics in Medicine and Biology*, 52:6651–6661, Oct. 2007.

[173] R. Richa, A. P. L. Bo, and P. Poignet. Motion Prediction for Tracking the Beating Heart. In *Proceeding of the 30th Annual International Conference of the IEEE on Engineering in Medicine and Biology Society (EMBS 2008)*, pages 3261–3264, Vancouver, BC, USA, Aug. 2008.

[174] R. Richa, A. P. L. Bó, and P. Poignet. Beating Heart Motion Prediction for Robust Visual Tracking. In *Proceedings of the 2010 IEEE International Conference on Robotics and Automation (ICRA 2010)*, pages 4579–4584, Anchorage, AK, May 2010.

[175] R. Richa, P. Poignet, and C. Liu. Deformable Motion Tracking of the Heart Surface. In *Proceedings of the 2008 IEEE/RSJ International Conference on Intelligent Robots and Systems (IROS 2008)*, pages 3997–4003, Nice, Sept. 2008.

[176] C. P. Robert and G. Casella. *Monte Carlo Statistical Methods.* Springer, 2004.

[177] K. Roberts. *Modellbasierte Herzbewegungsschätzung für robotergestützte Interventionen (Model-Based Heart Motion Estimation for Robot-Assisted Interventions)*. PhD thesis, Universität Karlsruhe (TH), 2009.

[178] K. Roberts and U. D. Hanebeck. Motion Estimation and Reconstruction of a Heart Surface by Means of 2D-/3D- Membrane Models. In *Proceedings of 21st International Congress and Exhibition on Computer Assisted Radiology and Surgery (CARS 2007)*, pages 243–245, Berlin, June 2007.

[179] J. Sainte-Marie, D. Chapelle, R. Cimrman, and M. Sorine. Modeling and Estimation of the Cardiac Electromechanical Activity. *Computers and Structures*, 84:1743–1759, Sept. 2006.

[180] T. A. Salerno, M. Ricci, G. D. Ancona, and J. Bergsland, editors. *Beating Heart Coronary Artery Surgery*. Wiley-Blackwell, 2001.

[181] M. Sauvée, A. Noce, P. Poignet, J. Triboulet, and E. Dombre. Three-Dimensional Heart Motion Estimation using Endoscopic Monocular Vision System: From Artificial Landmarks to Texture Analysis. *Biomedical Signal Processing and Control*, 2(3):199–207, Aug. 2007.

[182] M. Sauvée, P. Poignet, J. Triboulet, E. Dombre, E. Malis, and R. Demaria. 3D Heart Motion Estimation using Endoscopic Monocular Vision System. In *Proceedings of the 6th IFAC Symposium on Modeling and Control in Biomedical Systems (IFAC 2006)*, volume 6, pages 1–6, Reims, France, Sept. 2006.

[183] F. Sawo, V. Klumpp, and U. D. Hanebeck. Simultaneous State and Parameter Estimation of Distributed-Parameter Physical Systems based on Sliced Gaussian Mixture Filter. In *Proceedings of the 11th International Conference on Information Fusion (Fusion 2008)*, pages 1–8, Cologne, Germany, July 2008.

[184] W. E. Schiesser. *The Numerical Method of Lines: Integration of Partial Differential Equations*. Academic Press, 1991.

[185] F. C. Schweppe. *Uncertain dynamic systems*. Prentice-Hall, 1973.

[186] S. Sclaroff and J. Isidoro. Active Blobs: Region-Based, Deformable Appearance Models. *Computer Vision and Image Understanding*.

Special Issue on Nonrigid Image Registration, 89(2-3):1–29, Dec. 2003.

[187] M. Sehgal, K. Hirose, J. E. Reed, and J. A. Rumberger. Regional Left Ventricular Wall Thickness and Systolic Function during the First Year After Index Anterior Wall Myocardial Infarction: Serial Effects of Ventricular Remodeling. *International Journal of Cardiology*, 53: 45–54, 1996.

[188] M. Sermesant, P. Moireau, O. Camara, J. Sainte-Marie, R. Andriantsimiavona, R. Cimrman, D. Hill, D. Chapelle, and R. Razavi. Cardiac Function Estimation from MRI using a Heart Model and Data Assimilation: Advances and Difficulties. *Medical Image Analysis. Special Issue on Functional Imaging and Modelling of the Heart*, 10:642–656, June 2006.

[189] G. Shechter, C. Ozturk, J. R. Resar, and E. R. McVeigh. Respiratory Motion of the Heart From Free Breathing Coronary Angiograms. *IEEE Transactions Medical Imaging*, 23(8):1046–1056, Aug. 2004.

[190] R. D. H. Shinsuk Park and and D. F. Torchiana. Virtual Fixtures for Robotic Cardiac Surgery. In *Proceedings of the 4th International Conference on Medical Image Computing and Computer-Assisted Intervention (MICCAI 2001)*, volume 2208, pages 1419–1420, Utrecht, The Netherlands, Oct. 2001.

[191] R. M. Shoucri. The Pressure-Volume Relation in the Left Ventricle and the Pump Function of the Heart. *Annals of Biomedical Engineering*, 19:699–721, Jan. 1991.

[192] R. M. Shoucri. The End-Systolic Pressure-Volume Relation and Its Application to the Study of the Contractility of the Cardiac Muscle. In *Computers in Cardiology (CINC 2006)*, volume 33, pages 297–300, Valencia, Sept. 2006.

[193] W. Si, Z. Yuan, X. Liao, Z. Duan, Y. Ding, and J. Zhao. 3D Soft Tissue Warping Dynamics Simulation Based on Force Asynchronous Diffusion Model. *Computer Animation and Virtual Worlds*, 22(2-3): 251–259, Apr. 2011.

[194] D. Simon. Kalman Filtering. *Embedded Systems Programming*, pages 72–79, 2001.

[195] D. Simon. *Optimal State Estimation: Kalman, H-Infinity, and Nonlinear Approaches.* Wiley-Interscience, 2006.

[196] P. D. Spanos and R. G. Ghanem. Stochastic Finite Element Expansion for Random Media. *Journal of Engineering Mechanics of American Society of Civil Engineers (ASCE)*, 115(5):1035–1053, May 1989.

[197] D. Stoyanov, G. P. Mylonas, F. Deligianni, A. Darzi, and G. Z. Yang. Soft-Tissue Motion Tracking and Structure Estimation for Robotic Assisted MIS Procedures. In *Proceedings of the 8th International Conference on Medical Image Computing and Computer Assisted Interventions (MICCAI 2005)*, volume 2, pages 139–146, Palm Springs, CA, USA, Oct. 2005.

[198] D. Stoyanov and G. Yang. Stabilization of Image Motion for Robotic Assisted Beating Heart Surgery. In *Proccedings of the 10th International Conference on Medical Image Computing and Computer-Assisted Intervention (MICCAI 2007)*, volume 1, pages 417–424, Brisbane, Australia, Oct. 2007.

[199] A. D. Straw, K. Branson, T. R. Neumann, and M. H. Dickinson. Multi-Camera Real-Time Three-Dimensional Tracking of Multiple Flying Animals. *Journal of Royal Society Interface*, pages 1–16, 2010.

[200] M. Struwe. *Variational Methods. Applications to Nonlinear Partial Differential Equations and Hamiltonian Systems.* Springer, 2008.

[201] G. J. Subak-Sharpe, B. L. Zaret, M. Moser, and L. S. Cohen, editors. *Yale University School of Medicine Heart Book.* William Morrow and Company, Inc, 1992.

[202] B. Sudret and A. D. Kiureghian. Stochastic Finite Element Methods and Reliability. A State-of-the-Art Report. Technical Report UCB/SEMM–2000/08, Department of Civil and Environmental Engineering, University of California, Berkley, Nov. 2000.

[203] H. Suga and K. Sagawa. Mathematical Interrelationship between Instantaneous Ventricular Pressure-Volume Ratio and Myocardial Force-Velocity Relation. *Annals of Biomedical Engineering*, 1:160–181, 1972.

[204] T. Suzuki and T. Kanade. Measurement of Vehicle Motion and Orientation using Optical Flow. In *Proceedings of the 1999 IEEE/EEJ/JSAI International Conference on Intelligent Transportation Systems*, pages 25–30, Tokyo, Japan, Oct. 1999.

[205] T. Svoboda, D. Martinec, and T. Pajdla. A Convenient Multi-Camera Self-Calibration for Virtual Environments. *Presence: Teleoperators and Virtual Environments*, 14(4):1–26, Aug. 2005.

[206] R. Szeliski. *Computer Vision: Algorithms and Applications*. Texts in Computer Science. Springer, 2011.

[207] T. Takada. Weighted Integral Method in Stochastic Finite Element Analysis. *Probabilistic Engineering Mechanics*, 5(3):146–156, Sept. 1990.

[208] H. A. Talebi, K. Khorasani, and R. V. Patel. Tracking Control of a Flexible-Link Manipulator using Neural Networks: Experimental Results. *Journal Robotica*, 20(4):417–427, July 2002.

[209] D. Terzopoulos, J. Platt, A. Barr, and K. Fleischer. Elastically Deformable Models. In *Proceeding of the 14th Annual Conference on Computer Graphics and Interactive Techniques (SIGGRAPH 1987)*, volume 21, pages 205–214, July 1987.

[210] D. Terzopoulos and K. Waters. Analysis and Synthesis of Facial Image Sequences Using Physical and Anatomical Models. *IEEE Transactions on Pattern Analysis and Machine Intelligence*, 15(6): 569–579, June 1993.

[211] R. J. Theodore and A. Ghosal. Robust Control of Multilink Flexible Manipulators. *Mechanism and Machine Theory*, 38(4):367–377, 2003.

[212] L. H. Thomas. A Comparison of Stochastic and Direct Methods for the Solution of Some Special Problems. *Journal of the Operations Research Society of America*, 1(4):181–186, Aug. 1953.

[213] H. Timinger, S. Krueger, K. Dietmayer, and J. Borgert. Motion Compensated Coronary Interventional Navigation by means of Diaphragm Tracking and Elastic Motion Models. *Physics in Medicine and Biology*, 50(3):491–503, Feb. 2005.

[214] D. Uciński. *Optimal Measurement Methods for Distributed Parameter System Identification.* Systems and Control Series. CRC Press, 2005.

[215] Z. Ugray, L. Lasdon, J. C. Plummer, F. Glover, J. Kelly, and D. Rafael Martinec. Scatter Search and Local NLP Solvers: A Multistart Framework for Global Optimization. *Journal on Computing,* 19(3):328–340, July 2007.

[216] E. H. Vanmarcke. *Random Fields: Analysis and Synthesis.* MIT Press, 1998.

[217] E. H. Vanmarcke and M. Grigoriu. Stochastic Finite Element Analysis of Simple Beams. *Journal of the Engineering Mechanics,* 109 (5):1003–1214, Sept. 1983.

[218] M. Vořechovský. Simulation of Simply Cross Correlated Random Fields by Series Expansion Methods. *Structural Safety,* 30(4):337–363, July 2008.

[219] V. Vuskovic, M. Kauer, G. Szekely, and M. Reidy. Realistic Force Feedback for Virtual Reality Based Diagnostic Surgery Simulators. In *Proceedings of the 2000 IEEE International Conference on Robotics and Automation (ICRA 2000),* volume 2, pages 1592 – 1598, San Francisco, CA , USA, Apr. 2000.

[220] P. Walley. *Statistical Reasoning with Imprecise Probabilities,* volume 42 of *Monographs on Statistics and Applied Probability.* Chapman and Hall, 1991.

[221] C. G. Webster. *Sparse Grid Stochastic Collocation Techniques for the Numerical Solution of Partial Differential Equations with Random Input Data.* PhD thesis, The Florida State University, College of Arts and Sciences, 2007.

[222] G. Welch, B. D. Allen, A. Ilie, and G. Bishop. Measurement Sample Time Optimization for Human Motion Tracking/Capture Systems. In *Proceedings of Trends and Issues in Tracking for Virtual Environments, Workshop at the IEEE Virtual Reality,* pages 1–6, Charlotte, NC, USA, Mar. 2007.

[223] G. Welch and G. Bishop. An Introduction to the Kalman Filter. Technical Report TR95-041, The University of North Carolina at Chapel Hill, Apr. 2004. 95-041 pp.

[224] M. Wierzbicki and T. M. Peters. Determining Epicardial Surface Motion Using Elastic Registration: Towards Virtual Reality Guidance of Minimally Invasive Cardiac Interventions. In *In Proceeding of the International Conference on Medical Image Computing and Computer-Assisted Intervention (MICCAI 2003)*, volume 2878/2003 of *Lecture Notes in Computer Science*, pages 722–729, Montréal, Canada, Nov. 2003.

[225] B. R. Wilcox, A. C. Cook, and R. H. Anderson. *Surgical Anatomy of the Heart*. Cambridge University Press, 2004.

[226] G. Wolberg. Image Morphing: a Survey. *The Visual Computer*, 14: 360–372, 1998.

[227] F. Woolfe and A. Fitzgibbon. Shift-Invariant Dynamic Texture Recognition. In *Proceedings of the 9th European Conference on Computer Vision (ECCV 2006)*, volume 3952 / 2006 of *Lecture Notes in Computer Science*, pages 549–562, Graz, Austria, May 2006.

[228] S. Yue and D. Henrich. Manipulating Deformable Linear Objects: Attachable Adjustment-Motions for Vibration Reduction. *Journal of Robotic Systems*, 18(7):375–389, Feb. 2001.

[229] S. G. Yuen, S. B. Kesner, N. V. Vasilyev, P. J. D. Nido, and R. D. Howe. 3D Ultrasound-Guided Motion Compensation System for Beating Heart Mitral Valve Repair. In *Proceedings of the 11th International Conference on Medical Image Computing and Computer Assisted Interventions (MICCAI 2008)*, volume 1, pages 711–719, New York, USA, Sept. 2008.

[230] S. G. Yuen, D. T. Kettler, P. M. Novotny, R. D. Plowes, and R. D. Howe. Robotic Motion Compensation for Beating Heart Intracardiac Surgery. *International Journal of Robotics Research*, 28(10):1355–1372, Oct. 2009.

[231] S. G. Yuen, P. M. Novotny, and R. D. Howe. Quasiperiodic Predictive Filtering for Robot-Assisted Beating Heart Surgery. In *Proceeding of the 2008 IEEE International Conference on Robotics and*

Automation (ICRA 2008), pages 3875–3880, Pasadena, CA, USA, May 2008.

[232] S. G. Yuen, M. C. Yip, N. V. Vasilyev, D. P. Perrin, P. J. del Nido, and R. D. Howe. Robotic Force Stabilization for Beating Heart Intracardiac Surgery. In *Proceedings of the 12th International Conference on Medical Image Computing and Computer Assisted Interventions (MICCAI 2009)*, pages 1–8, London, UK, Sept. 2009.

[233] J. F. Zhang and Q. Y. Xu. Research on stability of periodic elastic motion of a flexible four bar crank rocker mechanism. *Journal of Sound and Vibration*, 274(1-2):39–51, July 2004.

[234] Y. Zhang, E. C. Prakash, and E. Sung. A New Physical Model with Multilayer Architecture for Facial Expression Animation Using Dynamic Adaptive Mesh. *IEEE Transactions on Visualization and Computer Graphics*, 10(3):339–352, June 2004.

[235] B. Zitová and J. Flusser. Image Registration Methods: a Survey. *Image and Vision Computing*, 21(11):977–1000, Oct. 2003.

Own Publications

[236] E. Bogatyrenko and U. D. Hanebeck. Adaptive Model-Based Visual Stabilization of Image Sequences. In *Proceedings of the 14th International Conference on Information Fusion (Fusion 2011)*, Chicago, Illinois, USA, July 2011.

[237] E. Bogatyrenko and U. D. Hanebeck. Visual Stabilization of a Beating Heart Motion by Model-Based Transformation of Image Sequences. In *Proceedings of the 2011 American Control Conference (ACC 2011)*, San Francisco, California, USA, June 2011.

[238] E. Bogatyrenko, B. Noack, and U. D. Hanebeck. Reliable Estimation of Heart Surface Motion under Stochastic and Unknown but Bounded Systematic Uncertainties. In *Proceedings of the 2010 IEEE/RSJ International Conference on Intelligent Robots and Systems (IROS 2010)*, Taipei, Taiwan, Oct. 2010.

[239] E. Bogatyrenko and U. D. Hanebeck. Simultaneous State and Parameter Estimation for Physics-Based Tracking of Heart Surface Motion. In *Proceedings of the 2010 IEEE International Conference on Multisensor Fusion and Integration for Intelligent Systems (MFI 2010)*, Salt Lake City, Utah, Sept. 2010.

[240] E. Bogatyrenko, P. Pompey, and U. D. Hanebeck. Efficient Physics-Based Tracking of Heart Surface Motion for Beating Heart Surgery Robotic Systems. *International Journal of Computer Assisted Radiology and Surgery (IJCARS 2010)*, 6(3):387–399, Aug. 2010. doi:10.1007/s11548-010-0517-5.

[241] E. Bogatyrenko, U. D. Hanebeck, and G. Szabo. Heart Surface Motion Estimation Framework for Robotic Surgery Employing Meshless Methods. In *Proceedings of the 2009 IEEE/RSJ International Conference on Intelligent Robots and Systems (IROS 2009)*, October 2009.